"Consciousness. Where does it come [...] from the human brain? Can the brain it[s] poses these and other intriguing questio[...] in philosophy and brain studies. In each [...] the heart of the struggle to explain subje[...] [...]ective, scientific terms." —*Science News*

"Succeeds in providing a very brief survey of the multitude of positions occupied by thinkers in this area. . . . The often quirky personalities and mannerisms of the interviewees shine through the text. . . . Blackmore herself comes across as spunky and clever, and the probing follow-up questions she occasionally asks prevent the interviews from seeming too repetitive or boring." —*Nature*

"Susan Blackmore posed the question 'What is consciousness?' to 21 leading scientists and philosophers who study consciousness for a living. It provokes all kinds of responses, ranging from jokes about psychedelic drugs to brow-furrowing discourses on life's meaning." —Richard Lipkin, *Scientific American*

"Are some scientists zombies? That is among the thoughts raised by this diverting collection of interviews with neurobiologists, philosophers, and others engaged in the study of the mind. . . . A very efficient overview of contemporary strands of thinking about its subject." —Steven Poole, *Guardian Unlimited*

"Blackmore interrogates 20 mind-body experts—philosophers, neuroscientists, psychologists, and various hybrids. She doesn't stand on ceremony, is persistent, probing, honest about her puzzlements, and happy to defend her own views if the occasion arises, which once or twice creates a bit of friction." —Tom Clark, Naturalism.org

"One remarkable aspect of the consciousness research field is the lack of agreement on what the key subject matter should be. What is the phenomenon for which we need an explanation? Susan Blackmore begins with these questions in *Conversations on Consciousness*, a collection of interviews with 21 prominent scientists and philosophers. Their answers introduce the reader to some of the concepts and puzzles at the center of this field. . . . The empirical and philosophical work summarized in the book is fascinating and easy to read." —Ephraim Glick, *EMBO Reports* (a publication of the European Molecular Biology Organization)

CONVERSATIONS
ON
CONSCIOUSNESS

What the Best Minds Think About the Brain, Free Will, and What It Means To Be Human

Susan Blackmore

OXFORD
UNIVERSITY PRESS

OXFORD
UNIVERSITY PRESS

Oxford University Press, Inc., publishes works that
further Oxford University's objective of excellence
in research, scholarship, and education.

Oxford   New York
Auckland   Cape Town   Dar es Salaam   Hong Kong   Karachi
Kuala Lumpur   Madrid   Melbourne   Mexico City   Nairobi
New Delhi   Shanghai   Taipei   Toronto

With offices in
Argentina   Austria   Brazil   Chile   Czech Republic   France   Greece
Guatemala   Hungary   Italy   Japan   Poland   Portugal   Singapore
South Korea   Switzerland   Thailand   Turkey   Ukraine   Vietnam

Copyright © 2006 by Susan Blackmore

First published by Oxford University Press, Inc., 2006
198 Madison Avenue, New York, NY 10016
www.oup.com

First issued as an Oxford University Press paperback, 2007
ISBN-13: 978-0-19-517959-0

Oxford is a registered trademark of Oxford University Press

The Library of Congress has cataloged the hardcover edition as follows:

Blackmore, Susan J., 1951—
Conversations on consciousness / Susan Blackmore.
  p. cm.
Interviews.
Includes index.

1. Consciousness.
2. Philosophers—Interviews.
3. Neuroscientists—Interviews.
I. Title.
B808.9.B53   2005
126—dc 22   2005019352

Contents

Acknowledgements

I would like to thank all the participants for making this book possible. Each one not only gave their time to the original conversation, but later read and checked the edited transcript and made corrections when necessary. I would like to thank John Byrne for encouraging me to record the conversations in the first place, Trudi Oasgood for typing the transcripts from the tapes, my daughter Emily Troscianko for preliminary editing, and my son Jolyon Troscianko for drawing the illustrations.

Although I spend most of my time alone at my desk, and think of myself as a loner, this book reminds me that I depend on you all.

All illustrations © Jolyon Troscianko, apart from the cartoon of Stuart Hameroff as a caterpillar which is reproduced from Journal of Consciousness Studies, 2 (1995), p.109, with permission © Imprint Academic, and the change blindness illustration which is © Kevin O'Regan.

Introduction

In the spring of 2000 I was preparing for a trip to Tucson, Arizona, for a conference called 'Toward a Science of Consciousness'. The first of these now-famous conferences had been held in 1996—and Stuart Hameroff and Dave Chalmers both tell stories about it. 'Tucson II', in 1998, had been bigger and already begun attracting a lot of attention, and I had been invited to take part in a plenary session on parapsychology. I had much enjoyed the whole event with its eclectic mix of neuroscientists, philosophers, and spiritual seekers. So I was now looking forward to the third, 'Tucson 2000'.

And I had an idea. I do a fair bit of work for BBC radio and television, and specially enjoy making radio programmes because of the freedom you get to express difficult ideas in depth. As the old joke goes: the pictures are better on the radio. So I contacted John Byrne, a producer I knew at BBC Bristol, and asked whether we might be able to make a programme for Radio 4 about consciousness. As it happened, our proposal never made it through the final stages of the complicated BBC selection process, but never mind. John lent me some broadcast quality recording equipment and I set off to Tucson to see if I could interview some of the great experts on consciousness that I knew would be there.

The process was great fun. It gave me a way of introducing myself properly to people I hardly knew, and an excuse for having in-depth conversations with old friends. I squeezed the interviews into gaps between the presentations, early in the morning, late at night, or during the one free afternoon; we did them in hotel rooms, in the plaza outside the conference hall, or out in the desert nearby. As we talked I came more and more to appreciate why the conference can only be called *Toward* a Science of Consciousness. There is so little agreement. And I learned such a lot—how feeble was my understanding of many of the theories I knew about; how different were some of the people when you got to ask them face-to-face what they really meant; how utterly confusing the whole field is. When the radio plan fell through I just wanted to keep going, and keep going I did. John kindly lent me the equipment again and I did the same at other conferences; at both the following Tucson events, and at two conferences of the Association for the Scientific Study of Consciousness, in Brussels and in Antwerp.

Eventually the idea of this book took shape. I realized that throughout the conversations I had been asking the same key questions, and there was almost no unanimity in the answers I received. These were the questions everyone was asking, and they lie at the heart of what it means to be human. I had had the good fortune to talk to some of the most famous names in the study of consciousness, and I could now share what I had learned by simply writing up my conversations.

As it turned out this was not as simple as it sounds. I thought it was important to let the people speak for themselves and not put my own spin on what they said—so I wanted to make the editing very light and keep as close as I possibly could to what they actually said. This meant doing the same with my own side of the conversations and sometimes I was horrified by how inarticulate I sounded. Even so, if I was keeping my conversationalists to their own words I would have to do the same with myself. So if you think some of my questions are inept then you know why.

But then I discovered that some people did not actually like what they had said. They wanted to rewrite their contributions in the style of a philosophy lecture or a neuroscience textbook. I resisted this very strongly; I urged them to let me keep their actual words, as spoken in the heat of a real live discussion in the desert or the lab or the hotel bar, as recorded on the tape. A few battles ensued which I did not enjoy. Some compromises had to be made, and I wish they had not. Almost always what people actually said was more fun, more lively, more interesting, and more daring than the words they wished to substitute. But when I really cared I stuck to my guns and the real words have remained in place. And if you want to know who argued about what I shall only say this—don't think you can guess because you are bound to be wrong. Just remember that, as close as I could get it, these are the real conversations that actually took place.

Once I had decided to do the book I realized that my collection of contributors was somewhat idiosyncratic, to say the least. Certainly if I had set out from the start to write a book called 'Conversations on Consciousness' I would have done it quite differently. I would have made a clear plan about the balance of people to invite, and would not have some of the glaring omissions you may have thought of yourself. For these omissions I can only apologize—both to the great minds I never conversed with, and to you the readers who might wish I had.

At the very end of the process I arranged a few last conversations. For one I have to thank Christof Koch for his kindness and quick intervention. I interviewed Christof at Tucson 2004 in April, in a cramped corner of the hotel with the cleaner audibly vacuuming nearby. When we had finished he asked me why I wasn't including Francis Crick. I explained that I would dearly love to but I knew that Francis was already 88 and unwell, and there was no way I would want to trouble him, even though I was, as it happened, going to a conference in San Diego a few days later. 'Then I'll ask him' said Christof 'I'm sure he'll say yes. He hates doing interviews about discovering DNA 50 years ago, but I know he'd enjoy your questions about consciousness.' And so it came about that a few days later Odile Crick warmly invited me to lunch and Francis and I spent a challenging hour battling over a topic dear to us both. Sadly this was the last interview Francis gave; he died in July 2004.

One final problem was a superficially trivial one—which order to put the conversations in. I tried making up groups or themes and got in a muddle; I tried working out which people introduced important ideas most simply so as to put them first, but got hopelessly bogged down. At one point I favoured a friend's delightful suggestion to order them by age. I could have started with Dave and his explication of the hard problem, and ended with Francis and his optimism for the future—or vice versa. But in between it made no sense, and in any case some people might have objected. So in the end I stuck with the very dull option of putting everyone in alphabetical order.

I asked everyone how they got into studying consciousness in the first place. This revealed some fascinating stories, from those who began in quite different careers, such as Dave, who began as a mathematician, Roger Penrose who is still a mathematician, Kevin O'Regan who studied physics, and Francis who began as a civil servant. I also asked them about their own work and their own particular theories. Some of these are very difficult to understand and some have always seemed to me to be daft. So it was wonderful to have the chance to ask the protagonists themselves what they really meant. You will see how I got on; in some cases I really did begin to understand but in others I remained just as perplexed as ever.

I did not start with stories about the past or with individual theories. Instead I began every conversation with the same question—what's the problem? I wanted to find out what it is about consciousness that makes people treat it as special or think of it as a problem that is different from other problems in science or philosophy. Of course some people, such as Pat Churchland, argue that it's not; that consciousness is just like any other scientific problem that needs to be solved by patient empirical work, and Kevin calls it a 'pseudo-problem'. But most people launched into versions of the mind-body problem or what Dave calls the hard problem. Briefly stated, the hard problem is the difficulty of understanding how physical processes in the brain can possibly give rise to subjective experiences. After all, objects in the physical world and subjective experiences of them seem to be two radically different kinds of thing: so how can one give rise to the other?

No one has an answer to this question, although some people seem to think they do, but asking it is worthwhile, if only for the depths of confusion it reveals. This confusion starts with the question itself and how best to word it. Dave himself originally worded it as I have done above, with the phrase 'give rise to'. He also talks about physical activity being 'accompanied by' subjective experience, implying a kind of dualism; in fact he defends a version of property dualism. But this might be completely the wrong way of thinking about the relationship between brain and consciousness. Perhaps, as the Churchlands argue, brain activity just *is* experience, or perhaps, as John Searle argues, brains *cause* experiences.

One thing that almost everybody agrees on is that classical dualism does not work; mind and body—brain and consciousness—cannot be two different substances. As Dan Dennett puts it 'there's no mystery stuff; dualism is hopeless.' Yet dualities of various kinds keep popping up all over the place, in spite of people's best efforts to avoid them. So I tried to winkle these out wherever I found them. Even saying 'give rise to' or 'generate' may imply that consciousness is something that is created by brain activity and therefore separate from it, which is why I challenged Susan Greenfield saying that 'the brain *generates* consciousness' and Richard Gregory that it 'generates sensations'; and presumably this is why Ned Block and Kevin refused to use the word 'generate'. I shall leave you to decide whether Susan really does avoid dualism by her temporary switch to 'correlations', whether Max Velmans succeeds with his reflexive monism or Vilayanur Ramachandran with his neutral monism, and whether Francisco's radical formulation really does escape the problem altogether. I cannot entirely decide for myself.

I am also unsure about the popular move from brains 'causing' or 'generating' consciousness to correlating with it; a move made not only by Susan but by Francis and Christof as well. In fact many people discuss the neural correlates of consciousness (NCCs)—meaning whatever is going on in a person's brain when they are having a conscious experience. This move sometimes appears to be the sensible and cautious strategy of considering correlations before going on to work out the underlying relationship, but sometimes it appears to be nothing more than a verbal trick designed to evade philosophical trouble. The lurking dualism can be sensed when people talk about NCCs as though the neural events are one kind of thing and the conscious experience is something completely different, and then imply that by moving from correlations to causes we can bridge the unbridgeable gap. Paul rejects both correlations and causality by insisting that experience just *is* a pattern of neural activation. And Kevin replaces it with the radical idea that experiences are not correlated with anything going on in the brain; rather they are what brains do.

Similar trouble can lurk in discussions of the difference between conscious and unconscious brain processes. For example, in answer to the first question, Bernie Baars asks what is the difference between knowledge that is conscious and knowledge that is unconscious, and answers in terms of Global Workspace Theory; Roger compares things that are conscious with things that are not; Ned compares information that is phenomenal with that which is not; and Christof compares neurons that give rise to consciousness with those that do not.

This distinction makes me very uneasy, and in these conversations I tried to explore why. A natural way of thinking about it seems to be something like this—we know that most of what goes on in the brain is unconscious; for example I am not aware of the way my visual cortex detects edges and corners or constructs 3-D shapes from the 2-D input; I am only aware of the tree I see outside my window: I am not aware of how my brain constructs grammatical sentences but only of the ideas I am trying to express and the words that come out of my mouth. So there must be an underlying difference in the brain between the conscious and unconscious processes.

But what could this mean? It might mean that although all brain activity is involved, there is some reason why we end up reporting experiences of trees and ideas, not neurons. Yet more often it is taken to mean that some brain cells or brain areas or types of neural activity or kinds of processing are the ones that create or give rise to or generate conscious experiences while the rest are not. This magic difference then throws us right back into the hard problem; for if we accept this difference we not only have to explain what it means for a physical brain to generate or produce consciousness, but why only some of its activity does so.

Finally I cannot leave this first question without mentioning the thorny topic of qualia. A quale is usually defined as the subjective quality of a sensory experience, such as the redness or sweet scent of a rose, or the rasping sound of a saw on wood. It is not the physical attributes of these things but the intrinsic property of the experience itself, and is private and ineffable. This philosophical concept has caused enormous trouble, and did so here. Many people mentioned qualia; indeed Francis, Rama, and Petra Stoerig began with them, then Dan Dennett denied their existence and Paul and Pat defended them, making things extremely confusing. It might help to say that if you take the definition of qualia very strictly then you have more or less committed yourself to the idea that experiences are intrinsically different from the physical world, and the hard problem is really hard. However, many people use the term much more loosely as a synonym for 'experience' and don't imply such a commitment. Watching out for this difference may help to avoid confusion.

All these interrelated issues can be summed up by asking where people stand on the following question—is consciousness something extra; is it something separate from the brain processes it depends on, or not? In a sense this is the central question that distinguishes the great theories of consciousness from each other. It has led to fierce arguments in the literature, and is important for many reasons. One reason is that, as neuroscience progresses and we learn more and more about the brain, we are gradually coming to understand such functions as vision, learning, memory, thinking, and emotions. So, when that understanding is complete, will there still be something left out—consciousness—that we haven't yet explained? Roger thinks so. So does Dave. He argues that when we have solved all the easy problems, there will still remain the hard problem of consciousness—a conclusion that is hotly denied by the Churchlands, Dan Dennett, and Francis. Dan has famously amassed what he calls 'the A team' to fight off Dave's 'B team' taunts of 'you've left something out'.

Once likened to a childish playground fight, arguments have long raged over the following question; when perception, memory and all other brain functions have been properly explained will there still be something left out? In an online debate about the importance – or fantasy – of a first person science of consciousness, Dennett declared himself leader of the "A team", with support from the Churchlands who are convinced there will be nothing more to explain, against Dave Chalmers' and John Searle's "B team", who are sure there will still be something left out – consciousness itself.

Another reason is that if consciousness is something separate then we may legitimately ask why we have it at all, or whether it has evolved for a purpose, because it would be possible for us to have evolved without it. In contrast, if consciousness is not something separate then these questions are plain daft. This is why I asked everyone for their views on zombies.

The philosopher's zombie is not some moulding half-corpse from Haiti that bumbles around in a trance; it is a thought experiment designed to help us think about consciousness. So ... imagine that there is a zombie Sue Blackmore. Zombie-Sue looks just like me, acts just like me, talks about her private experiences just as I do, and argues about consciousness just as I do; to anyone observing her from the outside she is completely indistinguishable from the original Sue. The difference is that she has no inner life and no conscious experiences; she is just a machine that produces words and behaviours while all is dark inside.

Could such a zombie-Sue exist? On the one hand, if you think that consciousness is something separate from the brain and its functions, then you would probably say yes. After all, it should be possible to take away that special consciousness (whatever it is) and leave all the other brain functions intact. The trouble is this leaves it as a total mystery why we should be conscious at all or what this extra 'something' could be or do. On the other hand, if you think that consciousness is nothing more than the functions of the brain, body and world, then you must deny that zombies could exist, because anything that could carry out all the usual functions of speaking, thinking, and acting would have to be conscious like you or me.

Put this way the answer 'no' seems preferable, yet the idea of zombies seems to have a life of its own. Even some functionalists, who should logically deny the possibility of zombies, find themselves imagining them. This is what Dan Dennett calls falling for the zombic hunch; giving in to the natural tendency to be able to *imagine* a zombie. So this is what I tried to explore in my conversations—were people just falling for the zombic hunch in spite of themselves, or did they really intend to defend their belief in the possibility of zombies? This is important because if they do hold this belief they must be thinking of consciousness as something that is separable from the brain and its functions. So note that I was very careful in how I worded my zombie question. I did not want to find out whether people could imagine a zombie—anyone can imagine a zombie—it is easy. I wanted to find out whether they really think that zombies could exist— in other words whether consciousness is separable from the physical person and its functions. Their answers were not always what I had expected. Some got into wonderful muddles and others just expressed their exasperation at the trouble the whole stupid zombie-thought-experiment has caused. Petra hates zombies, Francis said they're a contradiction in terms, and Francisco said the whole idea is absurd.

For fun I also asked many people whether they believed in life after death. I have long been interested in the fact that a personal life after death seems to me to be incompatible with a scientific understanding of the world, yet levels of belief continue to be high, especially in the United States. For example, a series of 1991 polls found that about 25% believe in life after death in such European countries as Britain, West Germany, Austria, and the Netherlands, with some Catholic countries having between 35 or 45%, and former communist countries much lower. Yet in the United States 55% believe in life after death. Not surprisingly most of my philosophers and scientists did not believe in survival; as Richard said 'one just snuffs out', but Stuart proposes a theory to explain it, and Kevin thinks that one day we will be able to download our personalities into computers and survive that way. But if I had expected definite answers from everyone, I was wrong, for several of my participants refused to be dogmatic about the issue.

'Do you think you have free will?' was the question that produced most diversity and personal agonizing. My intention was not to ask for a lecture on this grand old philosophical problem, but to get at a more intimate question—whether people believe that they personally have free will and how this belief (or lack of it) affects the way they live their lives. To be frank I had rather expected, before I began, that nearly everyone would intellectually reject the idea of free will while finding it hard to live their daily life without any such belief.

I say this for two reasons: first I have seen what students go through when confronted with philosophical arguments and scientific evidence concerning free will. They see that the whole system of brain and environment seems causally closed—in other words, that there is no room for an inner self or a conscious power to intervene—yet they go on finding it terribly hard to look on everything their bodies do as the product of prior events and their consequences. As Samuel Johnson put it so memorably 'All theory is against the freedom of the will; all experience is for it.' Some students just remain confused, while many say they decide to go on acting 'as if' there is free will even while not really believing in it.

Second, I have been through all of this myself. I long ago concluded that free will must be an illusion, and so over the years I have practised not believing in it. Eventually, with long practice, it becomes perfectly obvious that all the actions of this body are the consequences of prior events acting on a complex system; then the feeling of making free conscious decisions simply melts away. I had expected to find others who had gone through this somewhat disturbing change. Yet I was wrong. Everyone had something to say about free will, and many people had agonized about it. Dan Wegner and Pat both expressed the 'as if' option; yet, with the possible exception of Francis, no one completely rejected the notion of free will as I do, and no one seemed to share my experience of letting it go. Indeed Susan and John did not seem to believe me that it is possible to throw it off.

This was not the only question with which I tried to explore personal issues. I also asked people how studying consciousness had changed them as people, or changed the way they lived their lives. As Petra craftily surmised, I wondered whether people felt that studying consciousness had actually made them *more* conscious. For me, my scientific exploration of the nature of mind has been inextricable from my inner life and spiritual practice. I gave one example in talking about free will, but there are many others. One is the central issue of the nature of self.

What could a self be? The essence of consciousness is subjectivity, and subjective experience seems always to imply someone who is having the experience; in other words a self. But what sort of a thing could be the experiencer of experiences? And—even worse—what could such an experiencer correspond to in the brain? Rama, John and Francisco tackled the nature of self head on, and many others raised questions about it. Then there is the question whether one is the same self at different times. Thinking about this can be quite disturbing, and can begin to undermine one's natural sense of being someone. This is probably why questions such as 'who am I?' are used in some meditation traditions to bring about change.

I have certainly confronted such changes. I long ago concluded that there is no substantial or persistent self to be found in experience, let alone in the brain. I have become quite uncertain as to whether there really is anything it is like to be me. Yet, unlike with the illusion of free will, I have not (yet?) found that all sense of an experiencing self disappears. Although it does often depart, leaving only multiple experiences without anyone having them, the sense of 'me' tends to pop easily back into existence. So I was very interested to find out whether in this, or other ways, studying consciousness had changed people's sense of self or changed their consciousness or the way they live their lives.

Several people described their own experiences with meditation, drugs, and other altered states of consciousness; Stephen LaBerge talked about self-transformation through dreaming, and Thomas Metzinger and Francisco turned out to be long-term meditators, while others gave the impression that they would not be seen dead meditating. Several mentioned changes in their attitudes towards other creatures—both human and non-human animals—and others described how moral issues emerged from their study of consciousness.

I found it fascinating to hear how some people warmed to the question—for they had found their inner lives enriched by their work, or found themselves forced to integrate their intellectual and personal lives; for them inner work and intellectual work were inextricable, while others seemed quite happy to keep the two apart.

I learned an enormous amount from these wonderful conversations, and I thank everyone most sincerely for taking part. But do I now understand consciousness? I certainly understand the many theories about it a lot better than I did before, but as for consciousness itself—if there is such a thing—I am afraid not.

Bernard Baars

*Consciousness is a
real working theatre*

Bernie Baars was born in Amsterdam (1946), moved to Los Angeles when he was 11 years old, and studied psychology at UCLA. Rejecting the behaviourism of the time, he trained first in psycholinguistics, and then changed to cognitive neuroscience and became interested in artificial intelligence and consciousness. From the early 1980s he began developing Global Workspace Theory, which is described in his books *A Cognitive Theory of Consciousness* (1988) and *In the Theatre of Consciousness* (1997). He is Senior Fellow in Theoretical Neurobiology at the Neurosciences Institute in San Diego, California. He is co-editor of the journal *Consciousness and Cognition* and founding Editor of the web newsletter *Science and Consciousness Review* and of the Association for the Scientific Study of Consciousness (ASSC).

Sue **What, in your mind, is the problem of consciousness; what is it that makes it such a controversial area of science?**

Bernard In a way, it's funny that we need to ask that, because for all of written human history people have been fascinated by consciousness: in some sense it is one of the original fascinations of human thought.

If you ask questions about consciousness purely in terms of subjectivity—'What is it like to be you or me?'—you get into the classic mind-body paradoxes where you end up with the three classical positions in the mind-body problem: mentalism, physicalism, and dualism; and the dialogue—or rather, the dialogue of the deaf—on those particular issues, goes round and round and round and round and never gets resolved. So from my point of view the first thing that you must do if you would like to actually answer some questions, is pose the questions in a way that's answerable.

Sue **Go on then, pose me some questions in a way that's answerable. I totally understand this going round and round. Tell me a question we can ask to get us out of that.**

Bernard Well, here's the story that I would tell. The primary function of the nervous system, as far as we know, is to encode knowledge, to know things; and the technical term that's often used for this is representation. But there are unconscious forms of knowledge, or unconscious representations, and there are conscious representations. One of the clearly answerable questions that I think we have today is, what is the difference between two identical pieces of knowledge, one of which is conscious, and the other one is unconscious? That's an answerable question because it allows you to treat consciousness as a variable; and I would argue that anything in science that we can ask questions about has to be treated as a variable.

From that point of view the problem with the mind-body paradoxes is that they are always asked from one perspective, either from the inside perspective or the outside perspective; none of the classical positions allows us to ask about consciousness as a variable. A better question that William James asked in around 1890 is what happens if you put one kind of information, like a picture of a monkey's face, in your left eye, and another kind of information, like a picture of a sunburst, in your right eye? Well, that's called binocular rivalry, and it turns out that you cannot see both at the same time; one of them is conscious and the other one is unconscious. This allows us to compare them to each other; compare an unconscious representation to a conscious representation; and that allows you to ask testable questions.

In the last decade and a half we've seen many remarkable studies of binocular rivalry, so that now we know what the neurons are doing in the visual cortex; we know, apparently, at what point the neurons seem to recognize conscious events, and unconscious events; and we know how to ask those questions both in humans and in monkeys.

Sue So in binocular rivalry you have two pictures presented at once, and the experience alternates such that you seem to be consciously seeing first one and then the other. Then you measure what's going on in various parts of the nervous system. It sounds as though that ought to tell you what makes the percept conscious as opposed to unconscious—so what's the answer so far, from the research that's been done?

Bernard It's actually very nice. The brain regions for object recognition appear to be where the contents of consciousness emerge. There is a pathway from the eyes to the visual cortex. Below the cortex the pathway does not seem to involve consciousness. The visual cortex, in a very simplified way, can be thought of as a staircase: at the beginning of the staircase you have a map of your visual field with just very simple pixels, black and white dots; a little bit further on you have lines, and contrast edges between white lines and black lines; a little bit further on you have motion representation; and further on you have colour, and so on. At every step you add a little bit more analysis of the information that flows into your eyes. When you follow the staircase from visual region to region you finally come to object recognition cells in the bottom half of the temporal cortex, the cortex that is close to the temples of the head; and as you come to the end of the lower temporal cortex you finally come to the top of the staircase where you have object representation. And the best evidence that we have today—which comes from a dozen years of single-cell studies of all these different steps on the staircase—is that things become conscious on the top of the staircase, where you have cells that represent objects. Now that is over-simplified, but it's not a bad quick summary!

Sue But there seems still to be a mystery here to me, that what you're saying is that the difference between a perception that's unconscious and one that's conscious is a matter of which bit of the brain the processing is going on in. How can one bit of the brain with neurons firing in it be conscious, where another bit of the brain with very similar neurons firing in a very similar way is not? Don't we still have this explanatory gap?

Bernard There are lots of explanatory gaps. We are in the study of consciousness where Benjamin Franklin was in the study of electricity around 1800: he knew of a number of basic phenomena, and he might have known about the flow of electricity, and the usefulness of the stream metaphor—that things go from one place to the other, a little bit like the flow of water; that you can put resistors into the circuit, which are a little bit like dams, and so on. You have a useful analogy at that point in understanding electricity, which actually turns out to be not bad; but you have to improve it. So we're at a very primitive stage, but there are a few things that we can say.

Sue **And do we have around us now our Faraday, our Galvani—somebody who's going to sort out our understanding?**

Bernard We'll find out in about 100 years, but I think we have lots of people who would like to be the Galvani or the Faraday. There are some very nice proposals around, some of them coming from neurophysiologists. I'm a cognitive psychologist; I've made some proposals that I think work, at least in terms of the psychological phenomena; but I'm very much working on integrating my own theoretical ideas with neurobiology as well. Exactly how that's going to work is not entirely clear.

At this point there are some things that the psychological ideas explain, that the neurophysiology doesn't yet explain, and vice versa. For example, let's take the question that you ask—what makes one little patch of tissue in the brain a substrate for consciousness, and another little patch of tissue not a substrate for consciousness? There's a rather wonderful theory, proposed by Gerald Edelman, a Nobel prize-winner in immunology who has since become a neuroscientist. It's called 'Neural Darwinism' and is Darwinian in the sense that it deals with the cooperation and competition between massive numbers of neurons in the brain. What becomes conscious, in Edelman's view, is the winning coalition of neurons, those that outvote the other neurons. That is called the 'dynamic core hypothesis', and there's a great deal of evidence that's consistent with it.

Sue **You're best known for Global Workspace Theory, so I really would like you to explain what that means, in your own words, because I've known lots of people describe it, myself included, and you say we haven't got it right, we haven't understood it. So this is your chance!**

Bernard From my point of view, the metaphor that is useful for understanding consciousness is the theatre metaphor, which also happens to be quite ancient, going back at least to Plato in the West, and to the Vedanta scriptures in the East. The theatre metaphor, in a simple way, says that what's conscious is like the bright spot cast by a spotlight on to the stage of a theatre. What's unconscious is everything else: all the people sitting in the audience are unconscious components of the brain which get information from consciousness; and there are people sitting behind the scenes, the director and the playwright and so on, who are shaping the contents of consciousness, telling the actor in the light spot what to say. It's a very simple metaphor, but it turns out to be quite useful.

Sue **But some people think it's a very misleading metaphor. The way you've described it isn't quite the same as the way Dan Dennett describes it, but it has something in common with his idea of the Cartesian theatre; he says that although most people reject standard Cartesian dualism they still believe that there is something like a screen in the brain with someone watching it, some kind of mental theatre with me experiencing the show; and that this can't be true, because there is no place in the brain at which it all comes together; no top of a hierarchy of processing; no equivalent of the theatre or the audience. Now, I suppose you could say that it's only meant to be a metaphor; but some people would say we should throw it out because it's a completely misleading metaphor. What makes you say that it's a useful metaphor?**

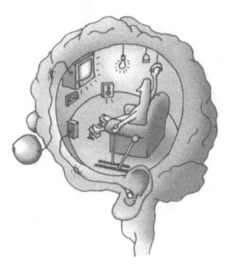

In *Consciousness Explained* Dan Dennett argues that although nearly everyone rejects Cartesian dualism, with its separate physical and mental stuffs, most people still think of consciousness as though there is a place or process in the brain where everything comes together and consciousness happens; as though there is a finishing line past which things become conscious and are displayed on the stage or screen to be appreciated by the inner audience of one. He calls this tempting, but false, way of thinking about consciousness Cartesian materialism, and discusses the give-away signs of being a Cartesian materialist.

Bernard   For one thing, Dan Dennett has changed his mind about my particular version of it; he basically acknowledges that there are versions of theatre metaphors, like my own and, I suppose, some other people's, which are not vulnerable to those particular criticisms. You don't have to have a little self sitting in the theatre; you don't have to have one point where it all comes together in the brain; you can have all kinds of more sophisticated ways of representing the information in the brain. In Newton's time people used the clockwork metaphor of the solar system, which is all wrong because there aren't long brass arms between the sun and the earth to keep us in orbit. Well, it's a metaphor! You have to use what works, and be very clear about the parts that are wrong. I should point out, by the way, that my work is based on very detailed computational models that work in reality, and which mimic human mental processes very nicely. The theatre metaphor is just a useful way of explaining it.

There are areas of convergence in the brain, and this top of the staircase I was talking about comes from research by a team led by Nikos Logothetis, a Greek American who is now in Germany, and who does binocular rivalry work using single cell recording in the macaque monkey. Logothetis finds that there is indeed a place in the visual system where 'it all comes together', this top of the staircase. It is the visual object recognition area I mentioned before. The staircase is also a metaphor of course, because it turns out that the top of the staircase cycles information back to every other step on the staircase, so this is not a simple staircase, but a very, very complex one. Furthermore, there is an engine underneath the staircase that keeps it all moving, called the thalamus, and all this stuff is necessary. If any of this machinery is damaged, consciousness is lost in a variety of different ways. Edelman talks about the thalamo-cortical system, which is really the best way to talk about the thalamus sitting underneath the cortex, making it all work, and which seems to provide the underlying dynamic system that allows one little piece of cortex to be at the bright spot on the stage for that particular moment. But it's a very dynamic system and can change from second to second.

Sue   So are you saying that information is coming in, and, in some kind of distributed process or neural network, is then made available to a whole lot of unconscious processes elsewhere in the brain?

Bernard   Exactly.

Sue   So in that case would it be right to say that whatever is being processed in that global workspace corresponds to the contents of consciousness?

Bernard   It's an interesting question, I'm going to evade it.

Sue   Oh no.

Bernard   I'm going to evade it explicitly, because there are certain things for which I think the evidence is good, and other things that are open; this is an open question to me.

Sue   When you talked about Edelman's theory, and Darwinian views, you said that when coalitions of neurons compete, the winner is the one that is conscious. It implies to me either that the non-winners are in their own way conscious but are over-shadowed in some way, or there must be some kind of switching on and off that makes them conscious when they win and not conscious when they don't, which doesn't seem to make sense. If we're talking about subjectivity what could it mean to switch it on or off? Which view do you take on that?

Bernard   Edelman argues that there is no external source of information other than the activation of the neurons themselves, so it's purely a vote, in terms of the mass of thalamo-cortical neurons working together in a single giant coalition. Other people argue, though, that there are cases where there may be isolated blobs of activation elsewhere in the nervous system. Of course the most famous example of that is the split brain studied by Roger Sperry and Michael Gazzaniga, where there really is no direct communication between the two hemispheres. I would argue that the evidence is all in favour of there being two consciousnesses in those patients: both hemispheres can, for example, answer questions; both of them can report perceptual experiences that are uniquely routed to those particular hemispheres; both have control over the hands and the fingers on the opposite side of the body. And so it would seem that both of them meet the criteria that we normally use for consciousness.

Sue   It's fascinating to think about this split brain question; people have given so many different answers to the question of whether there are two consciousnesses, one consciousness, many, none, whatever. But if you're going to take the view, as you do, that there are two consciousnesses in a split-brain patient, wouldn't it seem a small step to say that, because in an ordinary integrated person there is all sorts of activity going on in separate areas that's not necessarily connected to other areas, there are multiple consciousnesses in an ordinary person?

Bernard   That's an interesting question. You have to remember that in an ordinary person with an intact corpus callosum, there are at least 200 million fibres between the two hemispheres; they fire on average ten times a second; that creates two billion signals every second passing between the two hemispheres. In terms of Edelman's dynamic core hypothesis, that's enough to mobilize all these centres to work together on both sides of the brain. So their claim would be that if neurons are accessible to the dynamic core they will tend to go along with it. That is a hypothesis at this point, and it may be false; it is also conceivable that there are barriers to the flow of information between neurons elsewhere in the brain; that there are parts of the brain that are genuinely dissociated, as the expression goes. And of course there are famous examples about multiple personality disorder, where people apparently are profoundly dissociated when they move from one personality to another. So I think it's an open question.

Sue   What do you think happens to all this after death?

Bernard   I know of no evidence that consciousness remains. I realise that that's a painful thing; it's a painful thing to everybody; it would be a wonderful thing if one could believe in it—unless of course you believe in Hell—but I know of no evidence of continuity of self or consciousness after death.

Sue   Are you happy with that?

Bernard   No, I wish it weren't so; but one of the points that Freud makes about science is that science is always forcing people into believing things that they would rather not believe—and that goes back to the Copernican solar system. People were very upset about that; after Darwin's *Descent of Man* people were enormously upset; and rightly so—it's not that they were wrong to be upset. I think that one of the reasons why people have difficulty dealing with consciousness as a scientific issue is because it's terribly upsetting to many people that we don't have free will, that it's all due to these funny little cells firing in our heads, and all that sort of thing; and I sympathize with that. Some sort of godlike being, a platonic connection to the infinite, would be a rather wonderful thing to have; I just don't know of any evidence for it.

Sue   You've been promoting Global Workspace Theory for coming on for twenty years; how would you say it's faring in terms of the evidence accumulated during that time?

Bernard Well, the really exciting part is that the brain-imaging evidence is very, very strong by my lights—which is not to say that it proves the theory, of course; but it's highly consistent and very much unexpected by sceptics.

Sue And what about people's views of Global Workspace Theory?

Bernard It depends, as far as I can tell, very heavily on one's profession. The brain imagers think it's an interesting hypothesis; the psychologists have no idea what I'm talking about; the philosophers think it's all wrong because it doesn't explain subjectivity.

Sue Tell me how you got interested in consciousness in the first place.

Bernard I was born in Holland, and my family came to the United States in 1958, when I was 11 years old. I lived in Los Angeles, went to UCLA, and got interested in psychology. From day one at UCLA I began to realize that everybody was either behaviourist or trying to hide from others the fact that they were not behaviourist. Almost every professor I had an opportunity to talk to was going through the turmoil of the 'cognitive revolution'. They were dropping behaviourism, but being very cautious about it. There were still very powerful behaviourists in the department who thought cognitive psychology was all unscientific nonsense. I really started to wrestle with those issues from very early on and eventually evolved into the position of thinking that behaviourism is indeed all wrong—which I still believe now. One of my obsessions is this historical puzzle—why did perfectly good science get lost after 1900? In the nineteenth century psychology was preoccupied with consciousness, and later on, around 1900, 1910, it suddenly switched to behaviourism which involved radical rejection of everything that common-sensically we believe to be true, and which in fact *is* true.

Sue So you found yourself being educated in psychology in the midst of behaviourism. What was it for you personally that turned you against behaviourism?

Bernard I think it was the people I was influenced by who started to talk about meditation. I was interested in transcendental meditation at the time, and although the theory of transcendental meditation is an ancient Vedanta theory, and has the flaws of a theory that was probably produced about a couple of thousand years ago—subjectively, in terms of the experiences that people have, I suspect that it's reasonably accurate. There is something important there.

Very unfortunately there has not been much good research based on these very interesting phenomena; but the phenomena appear to be reported across many different cultures and times. And there is now a certain amount of reliable evidence that mantra meditation—where you repeat a word to yourself until it disappears—is associated with some distinctive brain effects, including high levels of alpha activity that spreads forward from the back of the brain.

What's hopeful about the last ten years is that we have this enormous improvement in instrumentation, so that we can look at the brain doing things without having to wait for the owner of the brain to die; so that we can see online what feelings people are having, whether they're feeling anxious or depressed, or when they're seeing things or hearing things, or whether they have intentions to do something. We now have, in effect, the brain-scope that has always been needed.

Sue Did your interest in meditation arise because you were practising it yourself, and if so, how much were you able to integrate what you learnt in meditation with the psychology and neuroscience that you knew?

Bernard Not very much at all. There turned out to be a great gap between what I appeared to experience in meditation, what my friends at the time also seemed to experience, and anything that we could explain. We did have theories but there was an apparent conflict. The organized meditation movement felt that it needed to control the evidence; they said they were interested in science, but they were really interested in science that served their own ends—and scientists, of course, are always coming to the wrong conclusion from the viewpoint of any orthodoxy. So, science was too messy for them, too unorthodox and uncontrolled; and I came to feel that although they had a lot of insights, they were not going to do the right scientific studies.

Sue That implies that you have some idea of what the right scientific studies are. If you could do anything you liked with scientific studies and meditation, what would you do?

Bernard There's a fantasy experiment I've wanted to do for a long time. According to the Upanishads, the Vedanta scriptures as they're called, the key notion is that there is a fourth state of consciousness. The first three are sleeping, dreaming, and waking; the fourth state is called pure consciousness. The definition of pure consciousness is very simple: it's consciousness without content. That doesn't sound particularly unscientific; it doesn't even sound particularly spectacular; it sounds fairly straightforward. So how could you assess consciousness without content?

One way to do it is to have people listening to a noisy air conditioner, for example, or a noisy heater, and have them do this meditation. If there are moments of consciousness without content, there should be gaps in the experience of external sound. That is after all the definition of pure consciousness. I used to notice gaps like that when I meditated. Now it's possible that I was just falling asleep, but if you put an EEG cap on people's heads you can see the classical slow waves of sleep when that happens. So you can rule those episodes out. It's also possible that people will give you false reports about gaps in the external noise level, because we know that their criteria are changeable, and that people may be motivated to have interesting experiences. But you can control for that also, by inserting artificial gaps in the noise source. People should report artificial gaps, as well as pure-consciousness gaps. So you can do a neat scientific study and use the very careful signal-detection methods that allow you to rule out false reports. Once you have that, then you've narrowed down the interesting moments in the meditation periods to a matter of seconds. If brain scans and EEG show distinctive brain signatures, you have something very solid.

There are also reliable reports of breath suspension, which turns out to be a good correlate of meditation. In these states of pure consciousness people have a spontaneous suspension of the breath that is not followed by what is called over-breathing, compensatory breathing—meaning people don't feel the need to take a deep breath immediately afterwards. That suggests that what may be happening is not a lack of oxygen—it's not like stopping breathing voluntarily at some point—but it's literally a brief period where the metabolic need for oxygen may be low.

Sue Isn't there a problem in getting people to report the cessation of content, in that trying to remember the task, or any way of reporting it, like speaking or pressing a button, will provide content and therefore destroy the very thing you're trying to report?

Bernard There are all kinds of interesting possibilities there. What you would need is a control group that hasn't been told to meditate, and ask—is there a difference between the control group and the meditation group?

It has to be said, by the way, that of the hundreds of experiments that have been done on meditation since the 1960s and 70s, very few of them are any good. Most of them have confounds in terms of expectation effects and placebo effects; they very frequently use highly committed people who spend years dedicated to this particular meditation method. And then you ask them, after doing it for six months, 'Do you feel any better?' And of course they're going to say yes. One way of getting around that is to look for physiological measures that people cannot fake, or simply do not know about. That's another good reason for using these quite wonderful brain-imaging techniques that we have these days.

Sue One of the fascinations for me of studying consciousness is that it's very difficult to make it separate from your everyday life. If you really ask questions such as, 'What is the nature of consciousness? What does it mean? What is it like to be me now?', you're forced into asking those questions in your very life, and therefore your life changes. Has this happened to you?

Bernard Yes. Most of my colleagues in cognitive psychology who came from a behaviouristic background used to deny that they were conscious of their own inner speech, and now I think they all hear themselves talking to themselves, as if a whole piece of their own inner experience has suddenly returned. The same thing is true about mental imagery: research on imagery used to ignore the question of consciousness, which seems kind of absurd. These days good scientists talk a lot about their imagery, their inner speech, moments of intention like the 'tip of the tongue', the nature of volitional acts, and so on. It's all stuff that William James would have been very comfortable with in 1890. So that's what can occur at a fairly obvious level.

Sue With the scientific study of subjective experience, there are going to be more people around who are altering their own experience, seeing more deeply into the nature of experience, even transforming themselves in some way. What do you think will be the consequences; are there going to be wider social implications?

Bernard My fantasy is that the famous split between the two cultures that C. P. Snow talked about in the 1950s will disappear. I would argue that one reason for the split in the twentieth century between the sciences and the humanities is that the sciences simply ignored all the wonderful things that the humanities were saying about consciousness, James Joyce being an example of that. Emotion is another topic that was neglected. Those two topics are coming back with amazing rapidity, and I think that within the next decade we'll see the end of the split, a kind of re-integration of a very divided century.

Sue **It seems to me that there's another kind of split in consciousness research, between the idea of investigating subjective experience as a scientific exercise for the sake of knowledge, and another, older approach, which has embedded in it the idea of transforming the self.**

This is a fascinating scientific enterprise—to study something which changes in the process of studying it; and untraditional within Western science to be concerned about the effect of doing the science upon the scientist. So we have here quite a challenge, I think, to ordinary science.

Bernard In fact, of course, science always changes our perception of reality. One of the impressive things about meditation traditions is that the reports are very widespread in different times and places. Similar experiences are reported by the Vedanta thinkers in the sixth century BCE, by Christian mystics a thousand years later, and by people today. If we did get a deeper scientific understanding of these processes, we might be able somehow to make it available to more interested people.

Sue **You know I sometimes wonder how feasible it would be to have a whole society of people who had been through such a transformation— who had meditated for years, who had let go of the conventional idea of self being separate from the world, and so on. It sounds like it ought to be a better society, but I'm not sure how it could work, and it might be impossible. What do you think?**

Bernard There's an old Zen saying that goes something like this: 'Before enlightenment chop wood, carry water; after enlightenment chop wood, carry water.' If that's the case, even after everybody's enlightened we will still be chopping wood and carrying water. It would be interesting to go to India and Nepal and simply observe those who are reputed to be advanced masters of meditation. Are they self-less in a way we could recognize? Do they curse at the fumbling novices who chop their toes instead of the wood? Are they willing to share and share alike? I'm a little sceptical about people who claim to be self-less—but maybe it's true!

Ned Block

*I'm trying
to refute
functionalism*

Ned Block (b. 1942) gained his PhD in philosophy from Harvard, held the Chair of the Philosophy Program at MIT, and since 1996 has been Professor of Philosophy and Psychology at NYU. He is best known for his criticisms of cognitive science and functionalism, for thought experiments such as the Chinese nation or China brain discussed here, and for his distinction between access consciousness and phenomenal consciousness. He edited *The Nature of Consciousness: Philosophical Debates* (1997).

Sue   What is the problem of consciousness?

Ned   The problem is, what *is* consciousness? More specifically, I'm interested in what consciousness is in the brain.

Sue   What do you mean by consciousness though? Why is that such a difficult and interesting problem for science or philosophy?

Ned   What I mean by consciousness, at least in this context, is the technicolour phenomenology; the 'what it's like'. Not everybody has that sense in mind; there are always different senses of consciousness; but that's the thing that's really interesting. Sometimes when people talk about consciousness they mean something about higher-order thought, or access, or monitoring, or self-reflection. Those look like the kind of thing we're making progress on in cognitive psychology, but what's really hard is something there's no progress on in cognitive psychology, namely the phenomenology. That's where the problem with the explanatory gap comes in—why is the neural basis of a certain phenomenal experience the neural basis of that rather than something else, or nothing?

Sue You are famous for making the distinction between access consciousness and phenomenal consciousness; can you explain what you mean by that distinction?

Ned Phenomenal consciousness is what I've just been talking about, that thing that we find so hard to understand how it could be a brain state, or how it could be supervened or determined by a brain state. Phenomenal consciousness is a thing such that we don't understand why it's determined by one brain state rather than another.

Access consciousness is what is often meant by consciousness—for example, I think it's what Freud meant by consciousness. When he talked about an unconscious state, he wasn't talking about something phenomenal, he was talking about something repressed, something you didn't have access to. This might, for example, be a vivid phenomenal state; somebody might have an image that it would be psychologically damaging for them to bring fully into the kind of awareness that underlies thought and reasoning; so they might have to repress that vivid image. It might be very phenomenal but it wouldn't be accessible.

Sue And you think these are two different things, do you? Do you think they'll remain two different things when we understand the brain even better?

Ned From what we know now, these seem different but highly linked things, but as we learn more we have conceptual improvement, and what's happened throughout the history of science is that concepts people start with, even very intuitive concepts, often split. In the seventeenth century, people didn't distinguish between heat and temperature. I was recently in Florence where, in the Museum of Science, they have the original devices; all the thermometers used by the Florentine experimenters in the very first systematic studies of heat and temperature. But they didn't know the difference between the two. So some of their methods measured heat and some measured temperature. For example, in one technique they would make a preparation, like a brick heated in a certain fire for a certain length of time, and then they would look at how much ice had melted in a certain period. So they found that some substances were hotter than another by that test—but they also had these weird, 400 unit thermometers, and by that test other things were hotter. So they were really trapped in a contradiction, because they didn't make this distinction between heat and temperature.

Sue You seem to be implying that the difference between access and phenomenal consciousness might be like that; that we need to make that distinction to get somewhere. Other people such as Dennett, and others too, would say that that's a false distinction and it will disappear. What do you think?

Ned I don't think it'll disappear but I think it might get more elaborated. Like for the heat question: it might even be that there are two kinds of phenomenal consciousness that we'll be able to distinguish on the basis of experiments, and then maybe we'll even be able to see it in our own phenomenology.

I learned recently something that probably a lot of people already know, which is that when you have a pain due to an injury, there are really two pains, a quick pain and a somewhat slower pain; and once I learned this, the next time I had a pain I could detect it. I think that phenomenology is not a static thing: the more you know the more you're likely to see in your own phenomenology.

There's a lot about phenomenology that's very obscure to us: for instance, do thoughts have phenomenology, or is it just the phenomenology of the words that are going through our head?

Sue You're implying here that studying consciousness, learning technical things about pain or the brain or anything else, actually changes your consciousness. So tell me, how have all these years of studying consciousness changed you or your life or your experience?

Ned Gosh, that's a hard question. Well, it's given me something exciting to think about. And yes, learning about wine changes what it's like to drink wine. So I don't see why learning something more general shouldn't change what it's like to do everything. I think I haven't learned much that has really changed my phenomenology, although the pain thing is one—but that's because we know so little.

Sue You made the distinction between phenomenal and access consciousness and I think you said we have learned nothing about phenomenal consciousness.

Ned Oh, I didn't mean to say that. But I do think we've learned very little about the scientific explanation of phenomenal consciousness.

It's something that we all have available to us on the basis of our experience, but as far as learning anything very serious about its nature goes, I think we don't know much.

Sue But some people disagree with you! For example, Paul Churchland says with respect to colour, which after all is one of the major issues, that once we really understand the whole colour space and how it's represented in the brain, we've done the job, we've understood the phenomenology. Then Kevin O'Regan says that if you take sensorimotor theory and you think about the mapping between action and perception, that explains everything that needs to be explained about experience. You disagree with both of those, presumably?

Ned I'm not sure that you have Kevin right. I think Churchland thinks that the mapping of colour space is important, but I doubt that he thinks it could completely explain everything; in Chalmers' terminology, he thinks that solving a lot of easy problems will add up to solving a hard problem; I don't think he thinks we've solved the hard problem.

To get back to Kevin O'Regan, I would describe him as not a phenomenal realist; not somebody who really believes in consciousness of the sort that I believe Churchland does. His theory is really a version of a behaviourist or a functionalist theory, and in my view it's not really based on data. I believe he had that theory long before there was any data. Most of the data he appeals to now wasn't around when he wrote his 1992 paper. He had the same views then; I'm sure he's had the same views all his life.

Sue But you could say that he's using his theory to predict things, and they've come true, and that that's what a good scientist should do.

Ned I think the view is an *a priori* view.

When I teach a class in the philosophy of mind, I usually start with the inverted spectrum. Some people talk about this as how things that look green to you look red to me and vice versa. I think there's a slightly different way of putting it that's better: that the things we both call red look to you the way the things we both call green look to me.

I don't think the words red and green should be thought to go with the experience, because I think that we might all be phenomenally different from each other, and that there's no one who has the real experience of red or of green. But the general idea is that even though we behave in the same way and our minds might even be organized in the same way, the fundamental underlying phenomenality of my experience of one colour might be like your experience of another colour.

So I go through this in my introductory classes, and about two thirds of the students usually say, 'Oh yeah, I see what you're talking about,' and some of them even say, 'Oh yeah, I've wondered about that since I was a kid'—in fact my own daughter, when she was seven, said, 'Oh, that explains why some people don't like purple, because they're not really experiencing purple the way I do when they are seeing purple; they get that experience when they see green or something.' But then about a third of people say, 'I don't know what you're talking about,' and I think that third is the group of people who, like Dennett or O'Regan, are one or another kind of functionalist or behaviourist; they're people who for some reason don't appreciate phenomenology and the difficult problems it raises.

Sue　**Do you mean they don't intellectually appreciate the problem of phenomenality, or do you mean that in some way their experience is so different that they can't appreciate it?**

Ned　I don't really know the explanation; it's something that I think would be wonderful to study. In fact Roger Shepard once suggested to me that he thought it was possibly some kind of defect in imagery—and I think it is something that could be empirically studied. You could get naive people and ask them various test questions like you would in a spectrum test and try to find what correlates with it. I don't think anybody's ever studied this, but I wouldn't be surprised if there is some difference in mental imagery between people whose reflex inclinations are to think that there's a problem of phenomenology and people who think that there isn't.

Sue　**You almost seem to be coming close to the idea that some of them would be zombies.**

Ned　Dennett often says that people say that maybe he is a zombie, but I don't know. I'm not saying they don't have phenomenology; I'm saying that there is some kind of failure of access to it, of the kind that would allow appreciation. Some component of mental imagery is perhaps my favourite hypothesis, although I've never tried to test it.

Sue　**Do you believe in the possibility of the philosopher's zombie?**

Ned　Well, there are two kinds of philosopher's zombies, so it's very important to distinguish between them.

Sue   Oh, I never knew that. Please distinguish away.

Ned   OK, I'll start with the one that's most intuitive, which is the person who is functionally like us, but physically so different that this person doesn't have the physical basis of phenomenology. For example, if you could make a person out of silicon chips...

Sue   or beer cans?

Ned   ... like the beer cans in Searle's example, or use a case that I used in a 1978 paper, the China brain. In fact this was the stimulus for John Searle's later Chinese room; he told me that he'd read my paper when he first gave his Chinese room paper.

Sue   Explain about the China brain.

Ned   OK, the idea is that you could assemble a group of people and have them communicate by satellite or by cell phone, so that each of them simulated what was in effect a neuron. They would interact by electronic means in a way that was like the way that neurons in the brain interact by electronic means.

I called it the China brain, because I said there are a billion people in China—not really as many neurons as there are in a brain, but something approaching that. They together would then control a body; all these people would be jointly the brain of that body, that robot. So the idea is that the robot, including its brain, might be functionally equivalent to a human being, in the sense that there are some corresponding states that interact with each other in a corresponding way. The question is whether the robot has phenomenology. Maybe there's no phenomenology, nobody home.

Sue   Is that what you think?

Ned   I don't say that I know that, because obviously I don't, and it's something for which scientific investigation is required. But if you believe in a neurological theory of consciousness you're going to be somewhat sceptical about whether this thing that is neurologically quite different from us, so different really in this extreme way, would have phenomenology. I think only a functionalist or a behaviourist, like Dennett say, would be sure that it does have phenomenology. So that's one kind of zombie, the zombie that's physically completely different from us, although functionally similar, with some set of corresponding states that interact in the same way and produce the same kind of behaviour.

Sue And what is the purpose of thinking that up; are you trying to refute functionalism?

Ned Yes, I'm trying to refute functionalism; that's what the purpose is.

Sue Now the second sort of zombie.

Ned The second sort of zombie is a creature that's physically exactly like us. This is the Chalmers zombie, so when Chalmers says that he believes in the conceivability and therefore the possibility of zombies, he's talking about that kind of a zombie.

My view is that no one who takes the biological basis of consciousness seriously should really believe in that kind of a zombie. I don't believe in the possibility of that zombie; I believe that the physiology of the human brain determines our phenomenology and so there couldn't be a creature like that, physically exactly like us, down to every molecule of the brain, just the same but nobody home, no phenomenology. That zombie I don't believe in but the functional zombie I do believe in.

Sue Do you believe you have free will?

Ned Yes. Well, maybe I should say yes and no, because I think that in many understandings of it, free will is a confusion. On the issue of phenomenology I'm completely different from Dennett, but on free will I'm almost exactly of his view. The trouble with free will is that it's both compatible and incompatible with determinism—and it's at once incompatible with determinism and incompatible with indeterminism. It's incompatible with determinism for the usual reasons; it's incompatible with indeterminism because chance alone doesn't make us free: if all of our actions happened by chance we wouldn't be free.

Sue So why don't you just say it's an illusion or it doesn't exist? Why do you agree with Dennett and say that it does?

Ned I don't myself think it really matters all that much which thing you say: you can say free will is a confusion and there's good reason for that; you can also say, 'Well, what do we really mean by free will? Well, what we mean is, I'm not in chains, nobody's pointing a gun at me, I could have done something different.' That's a kind of deflationary understanding of free will. We have an inflated conception of free will and a deflated conception. The inflated conception, where it means I'm somehow the author of my actions in a way that's not explicable by science, is a confusion. But if you take the deflated version of free will, where it just means I could have done something different, then yes, there is free will and it's compatible with determinism.

Sue And how does that play out in your life when you have to make a decision like where you're going to dinner after this, or whether you're going to tell me to stop now, and you need a drink? Do you feel that there's a little Ned Block inside there, who's responsible for making this decision and could do otherwise?

Ned No, no! I think I could do otherwise; but I don't think there's some little homunculus in there.

Sue So who is it who could do otherwise?

Ned Me, me, it's me; I'm the one who could do it.

Sue And who or what is that?

Ned It's a kind of constellation of states, an organized collection of states and their bases that interact with one another.

Sue And this 'you', this organized constellation—would you say it's this 'you' who has the experiences, has phenomenality?

Ned You see, part of my view is that I think there could be phenomenal states in us that aren't part of ourselves, that aren't integrated enough with the others to be thought of as a state of the self. This is one place where I differ from many other people who think about this.

Sue Ah, right. So let me try to get this clear.
Let's take the unconscious driving phenomenon, where you're driving along in the car, you're chatting to me, and you have such an interesting conversation with me that when you get to the car park and open your door, you don't remember the last ten minutes of driving at all. Clearly your body has been changing gears and so on—are you saying that there were conscious states associated with the driving, but they just weren't part of you, Ned?

Ned Well, there was a pilot study in a driving simulator, in which they got people to space out and then probed them, asking 'What are you experiencing now?' And people always report the last ten seconds or so. So I think that in those cases there is a moving window of memory. I don't think that's a case where you're having experiences but they're not you.

One example of that might be the extinction case. Extinction is a brain damage phenomenon in which if there's something on one side of space, usually the right side, people have no trouble seeing it, but if there's also something on the left side they can only see what's on the right. Nonetheless, as Geraint Rees has shown, the activation of the face-processing area in the brain which corresponds to that thing on the left is just as active as when they are seeing it.

Sue So you would say in that case that there was a conscious experience but it wasn't connected up to Ned?

Ned Yes. That's actually the best case I know of, because it's the only case I know of where the activation of the face area is just as strong as when the person does see it. For example, the binocular rivalry data showed that the shifts in the fusiform face area correspond in most circumstances to when people say they're having a percept as of a face. Yet in this extinction case you can have your face area activated just as strongly when you claim not to see something as when you claim to see it. I think that's a strong reason to believe that phenomenology of the face is going on in that brain, but isn't integrated into the rest of the person.

Sue But isn't the logical extension of what you've said something like this: here you and I are, sitting in this room, having a chat; most of your attention at the moment is on listening to my question; but we know that your brain is active in all kinds of ways—you'll be roughly monitoring visual things around, hearing sounds, prepared to respond if all sorts of things happen ... Are you saying that there are phenomenal experiences like that going on all the time which are not connected to you?

Ned Well, a lot of those things that are going on don't actually make the face area light up.

Sue So what's special here? Are you claiming that neural correlates have to be a particular area lighting up? And why on earth should activation of a particular subset of neurons in the brain, as opposed to all the rest of the brain, give rise to—I don't know what word you'd use—produce, generate, be associated with, be correlated with?

Ned I would not say generate—what I would say is determined.

Sue OK ... why on earth should this subset of neurons determine an experience while others don't?

Ned Well, that's the explanatory gap.

Sue So you're just happy to say 'I don't know'.

Ned Why does the state of the whole brain determine anything, determine any phenomenology? I think it's a fundamental mystery. Many people think that it's a mystery which will never be solved; other people like Kevin O'Regan, think it's a mystery which we have to solve by getting rid of the phenomenology: he thinks it's such a bad mystery that only by somehow analysing phenomenology away functionally can we come to terms with it. I think that's a short-sighted view. There have been many mysteries in the history of science—if nothing quite as bad as this, because after all, phenomenology is the hardest problem—and it can be useful to look back to the history of people's understanding of thought.

There was a time in the nineteenth century when people were terrifically puzzled by the same issue with respect to thought: how some kind of activations in the brain could possibly determine or constitute thinking—and I think now we've got a little further. One of the things we've done is to see that, in the case of thought, one shouldn't exactly be thinking about the brain in terms of activation of neurons; one should be thinking computationally. So now we have more ideas about how thought may work, and we think the computational approach is probably the right approach, and so people aren't so mystified by it; but we're just as mystified now, or even more mystified, about phenomenology. I think it's too early to throw in the towel and declare defeat.

Sue But in the case of thought, part of the progress has been because we've made machines that can do what we previously thought of as requiring some magical sort of thinking, like playing chess or solving problems or controlling things. But in the case of phenomenality, if we made such a creature, we wouldn't know whether it had experiences or not; so there's a big difference.

Ned Yes, there's a huge difference, and I think that's why the machine-oriented approach is hopeless when it comes to phenomenology.

Sue But you still think the analogy is valid in the sense that there have been what appeared to be insoluble problems that have been solved?

Ned Yes, but it's not just insoluble problems that have been solved. There are two features: first there are the cases where we didn't understand how the underlying basis of some mental phenomenon *could* be the underlying basis of it. And second, maybe even more important, is that it turns out we were looking in the wrong place, because we didn't have the computational concepts required to understand how thinking could work. I think the situation we're in is a little like what Tom Nagel described years ago in his famous paper on consciousness called 'What is it like to be a bat?' He used the analogy to a caveman: you tell a caveman that matter is energy, but the caveman doesn't have the concepts that would be required to understand that; and I think that we don't have the concepts required to understand how the mind-body problem could be solved. But I also think that those concepts are ones we wouldn't expect to have, and that we might get them in the future when neuroscience progresses further.

Sue Tell me how you got into all this in the first place.

Ned I think it was the inverted spectrum. When I was a college undergraduate that was the first thing that engaged me—I don't remember whether I first thought of it myself or somebody told me about it or what; I went to a course by Hilary Putnam on the philosophy of mind, and he may have mentioned it there. So that got me fascinated, and I've been hooked ever since, but I only really got interested in the science of it about ten years ago.

Sue You've made numerous contributions to the arguments about consciousness. Do you have a personal favourite, or one that you feel has been most valuable?

Ned Probably the Chinese brain one.

Sue Is that the same as the Chinese nation?

Ned The Chinese nation, yes.

Sue You've mentioned Dan Dennett's views, and clearly I'm far more enamoured of his destruction of the Cartesian theatre and his analysis of Cartesian materialism than you are...

Ned But nobody believes in Cartesian materialism, the idea that there's one place in the brain where consciousness happens; it was a straw man when he attacked it and it's still a straw man.

Sue People may not believe that it all comes together in one place in the brain, but lots of people talk about things coming into consciousness and going out of consciousness, as though it's a place—as though some information in the brain is 'in consciousness'.

Ned I talk that way, but what I mean is that there can be some information that's phenomenal and that that same information might exist in the brain in a non-phenomenal form.

Sue I think Dennett would call that Cartesian materialism, don't you?

Ned OK, if that's Cartesian materialism then I'm a Cartesian materialist, but the way he defines Cartesian materialism is in terms of a place. I think he really should have talked about a functionalized version of Cartesian materialism; that there's a functional place, a system, the system of consciousness of some kind.

Sue But he doesn't believe in a functional version any more than in a simple version, and most people do: most people believe that we can find the neural correlates of consciousness, the thing, or the place, or the system, or the united structure of neurons which correspond to what's in consciousness. But I would say, and I think he would too, that there's no such thing as being 'in consciousness'.

Ned That's because consciousness is so loaded with the place metaphor. I don't believe in the place metaphor, but I believe that there's such a thing as phenomenality and that it phenomenally can come and go. So I don't like the place metaphor—but you know, if anybody's an opponent of the Dennettian view it's me.

David Chalmers

*I'm conscious:
he's just a zombie*

Born in Australia (1966), David Chalmers originally intended to be a mathematician, but soon became interested in consciousness instead. He studied
at Oxford, before working in Douglas Hofstadter's research group for a PhD
in philosophy and cognitive science in 1983. His philosophical interests range
from artificial intelligence and computation to issues of meaning and possibility. He coined the term 'the hard problem', contrasting it with the 'easy
problems' of consciousness. After many years as Director of the Center for
Consciousness Studies at the University of Arizona, where he organized the
biennial 'Toward a Science of Consciousness' conferences, he has returned
to Australia as Director for Consciousness Studies at the Australian National
University in Canberra.

Sue **What's the problem? What makes consciousness such an interesting
and difficult thing to study?**

Dave   The heart of the science of consciousness is trying to understand
the first person perspective. When we look at the world from the perspective of science, we take the third person perspective. We see a
subject as a body with a brain, and with certain behaviour. We can be
terribly objective, but something very important about being a human
being is left out. As human beings we all know that it *feels* like something, from the inside. We have sensations, thoughts, and feelings.

You might say that there is this amazing movie which seems to be playing inside our mind—more wonderful than any movie you can actually go to in the theatres. It doesn't just have images and sounds. It has emotions and thoughts and the sensation of a body and all kinds of altered states which come around at different times. We all know this, and it's central to being a human being, but for some reason, in the last 50 or 100 years science has tended to ignore this.

Sue   **You can understand why can't you? It's very difficult to deal scientifically with the subjective experience of *feeling like me now* when it doesn't fit in at all with the study of neurons and brains.**

Dave   Sure, science is meant to be objective, and consciousness is subjective. So you might say that therefore science can't deal with consciousness. I think that's a fallacy.

A hundred years ago, psychology started as a science of consciousness. In fact the German psychologists conceived of what they were trying to explain in terms of a subject's internal conscious states. They developed detailed introspective methods, and collected data that way, but they descended into squabbles between different camps using different methods which yielded different conclusions. People got fed up with this because it seemed hard to settle the debates. Then, early in the twentieth century, the behaviourists took over. They said that from now on psychology is the study of human behaviour. Perhaps this made for a more rigorous and approachable kind of science. But many people feel that it is somehow like Hamlet without the Prince of Denmark. We are missing the central thing which we are trying to study.

So now I guess the question is how to bring consciousness back into the scientific world. My own attitude is that consciousness is data. As scientists we are used to talking about data and the results of certain measurements, and we try to build a science that deals with them. Usually these are objective data, but we have subjective data too. The data of consciousness—the way things seem to me right now—are data too. I am having a certain sensation of red with a certain shape right now. I am hearing a certain quality in the tone of my voice and so on. This is as undeniable as the objective data in the world of science. And science ought to be dealing with that.

Sue   **But isn't there a difference—an enormous gulf—between the subjective and the objective? Aren't they totally different kinds of thing?**

Dave   Yes, on the face of it they are enormously different things. So the question is, of course, one of the crucial questions in this field, 'How are we going to be able to explain subjective experiences in terms of the objective processes which are familiar from science? How do 100 billion neurons interacting in the brain somehow come together to produce this experience of a conscious mind with all its wonderful images and sounds?'

I think right now nobody knows the answer to that question. One could argue about whether such a reduction of subjective experience to a physical process is going to be possible at all. One thing that does seem likely is that we will find correlations. So when I have a certain colour sensation or a certain kind of emotion, there are going to be processes in the brain that go along with that kind of subjective experience. But that would be at the level of correlation. What we would eventually like is an explanation. That is, we would be able to look at the physical processes in the brain and say, 'Aha! Now I see why this gives rise to a subjective experience of this kind.' Right now nobody has a clue about that.

Sue **Do you have any sense of what such an explanation would look like? I mean an explanation of how one arose from the other that would satisfy you, and that you would say was more than just a correlation?**

Dave We do have analogies in other domains, of course. So when it comes to explaining the gene or explaining life, we have an explanation of what DNA molecules do—how they affect other processes in the body, how they lead to certain kinds of development, how they pass on information. Once we see that story we say, 'Aha! OK! That's all there is to being a gene. That explains what we needed to explain.' The question is whether we can do that for consciousness.

My own view is that we can't. Take the analogy with genes—what ultimately gets explained are the various different behaviours and functions which are associated with them. So you might say for consciousness, 'We'll explain the various behaviours and functions associated with consciousness. We'll explain how it is that my eye distinguishes and separates different sensory stimuli, how my brain integrates that information, how that leads to certain kinds of verbal reports and responses on my part.' But when it comes to consciousness those are the easy problems. Those aren't the central thing we are trying to explain. The hard problem is the question of explaining how it is that all this is accompanied by subjective experience. That seems to go beyond any mechanistic question about how the various behaviours and functions are produced.

Sue You have made an analogy here with trying to understand life. Some people say that consciousness is going to be just the same—that when we really understand all the mechanisms in the brain we'll understand consciousness. Why don't you think it's like that? Don't you think that if you went back, say 200 years, when people were talking about the *élan vital* and the life principal and what have you, they might have said just what you are saying now. 'I can't see how any understanding of chemistry inside a body would help me understand life—it's a different kind of thing.' Why isn't that a fair analogy?

Dave I think there is actually a disanalogy here, and it comes down to what really needs to be explained. When it comes down to explaining life, you say 'Well, what are the phenomena? What do we need to explain?' Biological beings reproduce, they metabolize energy from their environment, they use this in controlling their behaviour, they adapt and they grow. They compete with each other for resources. They evolve. All these are ultimately questions of behaviours and functions. What needs to be explained in each case are these matters of objective function.

Two hundred years ago the vitalists said, 'I can't see how you could have these behaviours, these functions, something as amazing as growth and reproduction. How could dead matter do that?' So they thought you needed to bring in a vital spirit. It eventually turned out that mechanisms could do all that, and so vitalism disappeared. But what's interesting is that this shows what even the vitalists conceded, that when it came to explaining life, all we needed to explain were objective third person behaviours.

Now with consciousness, things are completely different. We can all agree on what needs to be explained. There's my behaviour and my responses and my reports, sure. And let's all concede, at least for the sake of argument, that science might be able to explain those. The trouble is that we haven't exhausted what needs to be explained. We've left out the central datum; the datum of subjective experience. And that seems to have no analogy in the life case.

Sue But wait a minute. Aren't some of these 'data of subjective experience' turning out to be illusory? For example, there's the feeling that consciousness *does* something. This is a very ordinary human experience, in which it seems to me that I *consciously* decide to do something and then it happens. And yet there are many scientists who say 'Well, actually that's an illusion. These decisions are made, the body acts, but consciousness doesn't have any role.'

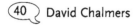

What happens when an animal or person dies? Something seems to have departed—something like a vital spark that makes the difference between life and death. In the nineteenth century philosophers believed that there really was such a thing and called it the *élan vital*, or vital spirit. But when twentieth century science began to unravel the mysteries of how living things work and reproduce, the idea was abandoned and people now accept that there is nothing more to being alive than complex, interrelated, biological functions. Is consciousness going to go the same way? That is, once we understand all the functions of thinking, perceiving and remembering will we realize that there is nothing left to call 'consciousness'. Dave and Stuart say no, while many others are convinced it will.

Couldn't it be that all these feelings about what conscious experience is, or what subjectivity is, will eventually just disappear, and we'll see them all to be some kind of illusion?

Dave I wouldn't want to say that people are infallible about the contents of their consciousness, because clearly that's false. For example, you might put an ice cube on my back when I was expecting a match. I could think for a moment that I am having a sensation of hot, but then after a moment I realize that no, actually that was a sensation of cold. But it's one thing to say that we can be mistaken about certain subtle things in the fringes, but could I really be mistaken about the fact that right now I am having a visual experience; a visual image with certain shapes and colours and so on? I think that is just impossible.

Maybe I am wrong about certain subtle features of the image. Maybe, for example, I think there is more going on in the background of my visual image than there really is. But to say 'Well, maybe I'm not really conscious at all,' that seems to be going too far. Descartes, of course, said that this was the one thing we know more certainly than anything else. '*Cogito ergo sum*: I think, therefore I am.' What he was really talking about was consciousness.

Sue And do you agree with Descartes?

Dave I do agree with Descartes on that. There is no doubting that we are conscious. I think we can only *doubt* that we have consciousness in philosophical moments—when philosophers are arguing about this and they say, 'Maybe it will turn out that consciousness doesn't exist.' But I think that this is simply going contrary to the manifest data of subjective experience.

Sue You talked earlier about the 'easy problems' and the 'hard problem', and this distinction is probably what you are most famous for. In fact, *everybody* now seems to start any discussion of consciousness with an account of the 'hard problem'. Can you tell me how you came to categorize it that way?

Dave I never thought of this as a terribly profound distinction to make. I thought I was just stating the obvious. I gave a paper at the first Tucson conference on consciousness, back in '94, and early in the conference I got up and wanted to say some substantive things about consciousness. So I thought, 'OK, I'll start by stating the obvious—what needs to be explained is behaviour (those are the easy problems), and subjective experience (that's the hard problem).' Now this was meant to be just the prelude before I went on to say something more profound.

    Of course, what everybody remembers are those first five minutes at the beginning. I guess it turned out to be useful for the field to have a short tag for the problem. But now it's taken on a life of its own. I don't think I added anything profound and original, because everybody who really thinks about consciousness knows that the hard problem is the problem of subjective experience, and they have known this for hundreds of years.

Sue You have described the hard problem as the difficulty of explaining how subjective experience arises from an objective world. Is this the same as the mind-body problem? Is it the same as the problem that leads to Cartesian dualism? Or is it a different problem.

Dave I think it's in the same ball park. The term 'mind-body problem' covers a multitude of sins. One is this question: 'How is it that the brain can support subjective experiences?' Another one is: 'How can the brain support thought, or rationality and intelligence?' Maybe that is not quite the same problem, because it's closer to the domain of behaviour. Another question is: 'How can the mind affect the physical world?' That's very closely related. But they are slightly different problems. We can think of the hard problem as the real core of the mind-body problem.

Sue   And now to those profound bits—what's your own way of tackling the hard problem?

Dave   I am not going to sit here and tell you that I'm now going to say something profound, and then say it!! But, OK, I think there are reasons, which I have touched on, for saying that subjective experience can't be reduced to a brain process. No explanation solely in terms of brain processes will be such that we can deduce the existence of consciousness from it. I think someone could know all the physical facts about the world and still not know about consciousness. So if the relationship between brain processes and conscious experience isn't one of reduction, what is it? Obviously there is going to be a very close correlation and a connection. What a science of consciousness needs to do is to systematize that bridge.

This raises deep questions of metaphysics. What is there in the world? What are the basic components of the world? In physics this happens all the time. Nobody tries to explain, say space or time in terms of something which is more basic than space or time. It's the same with mass or charge. They end up taking *something* as fundamental. My own view is that to be consistent we have to say the same thing about consciousness. If it turns out that the facts about consciousness can't be derived from the fundamental physical properties we already have, like space and time and mass and charge, then the consistent thing to say is, 'OK, then consciousness isn't to be reduced. It's irreducible. It's fundamental. It's a basic feature of the world.'

So what we have to do when it comes to consciousness is admit it as a fundamental feature of the world—as irreducible as space and time. Then we need to look at the laws that govern it, at the connection between the first person data of subjective experience and the third person objective physical properties. Eventually we may come up with a set of fundamental laws governing that connection, which are akin to the simple fundamental laws that we find in physics.

Sue   I understand that you want to try out the idea that consciousness is a fundamental principle of the universe, but you were talking there about correlations. Most people, when they talk about the 'neural correlates of consciousness', mean that they take one thing (such as a subjective report)—and another thing (such as something they can measure in

the brain)—and try to see if they are correlated. Now if you were just saying that, it wouldn't help would it? I take you to be saying something more fundamental than that—that consciousness is not just one more thing that can be correlated, but that it underlies the world in some way, or that it forms a framework.

You made an analogy with space and time, and space and time in physics are basic principles, used to structure everything else. So if you were going to make that analogy work you would have to say something similar about consciousness. Is that what you are trying to do, and can you do it?

Dave   I am not saying that consciousness structures everything else in the world. All I am saying here is that it is a fundamental feature of the world. The question is how can we get to a theory? How can we have something that looks like an explanation of consciousness when we just have these subjective phenomena and these physical processes in the brain? If all we have as our fundamentals is, say, space and time and mass, then consciousness isn't even going to get in to the picture. So we put consciousness in to the picture and we study the correlations.

In this picture, everything that's going on in the study of the neural correlates of consciousness will turn out to be important work. You might say it's going to be even more important, because by studying the correlations between the first person and the third person we are gradually moving towards those fundamental principles which bridge the divide.

Sue   If consciousness is somehow that fundamental a principle, wouldn't you expect it to be ubiquitous? Are you coming close to a panpsychic view here, where everything is conscious?

Dave   I think the view that consciousness is irreducible is neutral in the question of whether consciousness is ubiquitous. You could say that it is irreducible but rare. I mean some fundamental properties are rare. There are huge areas of vacuum throughout space in which there is no mass, for example. So maybe there are huge areas in which there is no consciousness.

It is true, though, that it is natural to speculate. After all, it is very hard to draw the line for where consciousness stops. We think people are conscious, almost all of us think chimps, dogs, and cats are conscious. When it comes to fish and mice, some people might deny it. But fish and mice have perceptual fields and it's plausible that they have some kind of conscious experience. Then you just go further and further down.

My own view is that where you have complex information processing you find complex consciousness. As the information processing gets simpler and simpler you find some kind of simpler consciousness.

Sue   This would lead to a very odd thought though. You say that associated with all kinds of information processing is some kind of consciousness. In a human being there may be multiple sorts of information processing going on at once—I mean different bits of our brain are doing all these different clever things—and only some of them are what we would call 'my consciousness'. It seems to me you must be saying that there are _multiple consciousnesses_ which I don't know about going on in this brain here.

Dave   Well—this raises some interesting questions about the self and the subject. This is only speculation, but on a panpsychic view I would imagine that the kind of consciousness that you would find throughout most of the world is incredibly simple and undifferentiated and not very interesting. Some of the time that basic field of consciousness might come together into unified, coherent, bounded objects that we think of as selves. Now what the conditions are for that, I think nobody knows. Maybe it's got to do with certain kinds of very systematic, coherent information processing. So that means that in the vicinity of my brain there's this one remarkably coherent system of information processing which corresponds to 'me'. Now, as you say, there are other things going on in my body, and one would have to say that there are experiences associated with those. But those don't give rise to selves or to subjects, and they have nothing to do with me.

Sue   So would they be more like the sort of consciousness in an animal that had no concept of self?

Dave   Or maybe even simpler. Let's look at an incredibly simple system like a thermostat. Who knows? Is a thermostat conscious? It would only be speculation, but just say it was. It would at best be a tremendously simple and primitive form of consciousness. One state here, another state there, but nothing corresponding to what we would think of as thinking, or intelligence, or a self.

Sue   You're touching here on one of those other problems that has become central in arguments about consciousness. That is, whether a system

carrying out some intelligent behaviour would necessarily be conscious by virtue of doing that behaviour. And this comes close to your zombie theory. Would you like to explain about zombies?

Dave Sure. I think in the actual world, intelligent behaviour and consciousness very likely go together. So when you find a system which is behaving like me and talking like me—it's probably conscious. But it seems that I could *imagine* a system which was behaviourally just like me, it walked and talked just like me, it got around its environment, but it didn't have subjective experience at all. Everything was dark inside. This would be what philosophers like to call a zombie—a being entirely lacking consciousness.

Now such a being would be tremendously sophisticated. You couldn't tell the difference from the *outside*, but there would be nobody home *inside*. Here I am sitting talking to you. All I have access to is your behaviour. Now you seem like a reasonably intelligent human being, you're saying articulate things that suggest a conscious being inside. But of course, the age old problem is 'How do I know?' It's at least logically consistent with my evidence that you are a zombie.

Now I don't think you are, but the very logical possibility of zombies is interesting because then we can raise the question 'Why are we not zombies?' There could have been a universe of zombies. Think about God creating the world. It seems logically within God's powers (and of course the use of 'God' here is just a metaphor) to create a world which was physically just like this one with a lot of particles and complex systems behaving in complex ways, but these were just androids. There was no consciousness at all.

And yet there *is* consciousness. So that's been used by some people, including me, to suggest that the existence of consciousness on our world is a further deeper property of the world than its mere physical constitution.

Sue So are you saying that you believe such philosopher's zombies are possible and the fact that we have consciousness means that we have to add something to the explanation?

Dave I think they're probably not possible in the sense that no such thing could ever exist in this world. I think that even a computer which has really complex intelligent behaviour and functioning would probably be conscious. What is interesting though, is that it doesn't seem contradictory to suppose, at least in the imagination, that someone, somewhere, in some possible world could behave like me without consciousness. But our world isn't like that. So that's an interesting fact about our world!

Sue You say our world isn't like that. Does this make you a functionalist? Are you saying that, in our world, anything that carries out a certain function must necessarily be conscious?

Dave In some very broad sense I am a functionalist. I think that behaviour, and function, and consciousness go together. They are very tightly correlated and associated. But I am not a functionalist in the strong sense of saying that all there is to consciousness is the functioning. Some people say that all we have to worry about is functioning and the behaviour and the talking. I think that is just manifestly false because of the direct data of subjective experience. We have correlation of the two without any kind of reduction of one to the other.

Sue I want to get this absolutely clear because people talk about your views on zombies a lot. You are saying that logically you can conceive of a world in which there would be intelligent, behaving creatures who went around saying things like 'I am conscious' and 'I'm experiencing red right now' and so on, but didn't have any subjective experience. But you think that in this real world we are in that's not possible and anything that does these behaviours will necessarily be conscious.

Dave That's exactly right.

Sue Good!

Now it seems to me that the zombie question is related in an interesting way to the question of evolution—that is, 'Has consciousness evolved for a reason?' Because if zombies were possible in this world then you would have to explain why we aren't zombies. You would have to say 'We *are* conscious, so there must be some function for consciousness, or some reason why evolution added on consciousness.' Whereas if you take your view, that necessarily any system that does all these things must be conscious, then there is no necessity that evolution has produced consciousness for a reason is there?

Dave Not necessarily, no. On my view, of course, evolution is going to select physical systems for their physical functioning. Once you have a system which functions like that it will be conscious. So therefore consciousness will evolve. But did that system evolve because it was conscious? Was consciousness doing something for that system? I think right now, nobody has any answer to that question.

People put forward speculation—maybe the function of consciousness is planning or decision making or integrating information or whatever. But then as soon as such a hypothesis is put forward the questions just get raised 'Why couldn't that have been done without consciousness? Why couldn't you just have had these brain processes which produced that conclusion with no subjective experience anywhere?' And of course you can use zombies to illustrate this point. You can imagine, at least hypothetically, that zombies could have existed which did the kind of things that we do but without consciousness. Now of course in our world consciousness *is* here so that is the difference between us and zombies. It does raise the very deep question of what consciousness is for.

One possibility is that consciousness is a non-physical thing that interacts with the physical world, as Descartes thought. It could then be selected for by virtue of its actions. That's regarded as somewhat implausible though, because it comes into tension with our view of the physical world as revealed by physics. Although in turn some people think there is room for it in quantum mechanics.

Maybe there is another way of approaching the question 'Why is there consciousness?' You might say that consciousness is a thing which gives our lives meaning. It makes our lives comprehensible and interesting and a locus of value. And in a world of zombies there would be no meaning.

Sue You mentioned quantum mechanical approaches to consciousness. Do you think these are valuable?

Dave I think they are interesting but extremely speculative. One basic problem is this. In classical neuroscience you may have 40-Hz oscillations in the brain, or various interactions, but why should any of that give you consciousness? People can't see how. So they say, 'Ah—we need something new. Something extra. An extra ingredient. Let's say it's a collapsing quantum wave function in our microtubules.' But now the question comes up again. But why should collapsing wave functions in microtubules give you consciousness? You're not really any closer.

Sue Do you think you have free will?

Dave I don't know, I really don't know. And the reason I don't know is that I don't know what it means to have free will.

I know that most of the time when I want to do something I do it, and most of the time that seems good enough. If I want to go down to the grocery store, I can go to the grocery store, except if somebody is locking me up in prison then I can't. But I can, so I am free.

Now someone is going to come back and say, 'Aha, but what you *want* to do, the fact that you want to go to the grocery store, that was determined all along, and therefore you are not free.' And there are moments when I actually think, 'Well, that worries me. I can't choose what I want, because that is already determined.' But then I just say, 'Well, how else could it be?' Who would want to be able to choose what they want? That is just part of who I am. So maybe this further kind of free will, where one can choose who one is going to be and what one is going to want in some undetermined way, is just an illusory desire and would at the end of the day be useless, because this is who I am.

Sue Do you feel that your life has been changed by all these years of thinking about consciousness?

Dave I think it would be nice if the answer were to be 'yes'. I think it affects little things. For a while I was very tempted to become a vegetarian because I didn't want to eat anything which is conscious. Then I started to develop views about consciousness which suggested that it wasn't just cows and pigs and so on which were conscious. Now, if I had still stuck to my principles, I was going to go very hungry. So I said, 'OK, what this suggests is that it's not consciousness that matters, it's complex consciousness that's morally and ethically significant.' So the consequence is that I don't mind eating fish and perhaps chicken and certain simple organisms. I have some qualms there, but I am not totally uncomfortable eating meat, which is probably convenient because as a human being I like the taste quite a lot.

Sue How did you get in to all this in the first place? Have you worried about consciousness ever since you were a kid? Or was there something particular that started you thinking about it?

Dave I do know that when I was ten I discovered I was short-sighted. It turned out that I had one very good eye, but the other was very blurry, and one day I got glasses that gave me binocular vision. Now the world wasn't just sharp, it was also deep. And I wondered, 'How does just getting glasses suddenly make the world feel deep?' I could understand it from the third person point of view, but not from the first person point of view.

Later on, as an undergraduate studying mathematics and physics, I used to sit around the table talking about consciousness all the time with my friends. I thought it was way too much fun that one could actually make it one's profession. It seemed kind of illicit, somehow.

I still think that from time to time. I would have loved to have been a mathematician or a physicist 500 years ago, at the time of Newton when nobody knew anything. That would have been exciting! So many open frontiers! Mathematics and physics are still very interesting, but there is a sense that we've got the basic framework and are filling in the gaps. I wanted to be on one of those frontiers. I was just obsessed at this point by the problem of consciousness, so I made a leap of faith. I got out of mathematics and physics and started trying to turn my wild ideas about consciousness in to something vaguely down to earth within the context of being a philosopher and a cognitive scientist. And in the end it seems to have more or less worked out, but that's not to say that anyone is ever great at doing philosophy. It's just too hard.

Sue I bet you are glad, now, that you had the courage.

Dave Yeah. You know I have to say it took me a while to work up the courage. For the first year or two while I was talking about doing this, everybody said I was crazy. My family said I was crazy. They said, 'You're pretty good at mathematics. What's all this nonsense about philosophy? Nobody gets anywhere doing philosophy.' But I think it's turned out that I have a more interesting life this way than I ever would have as a mathematician.

Sue What do you think happens to consciousness after death?

Dave I don't know for sure. But I'm inclined to think that my consciousness ceases to exist. Whether or not consciousness is reducible to the brain, my consciousness seems to depend on my brain. Damage my brain, and you damage my consciousness. After death, my brain will disintegrate, so my consciousness will disintegrate too. If a panpsychist view is true, it could be that corresponding to my disintegrated brain will be some disintegrated fragments of consciousness. But I don't think these fragments would count as my consciousness in any recognizable sense. I'll probably cease to exist. Then again, no one understands consciousness, so I could be completely wrong. That would be nice!

Patricia & Paul Churchland

The brain is a causal machine

The visual sensation of redness is a particular pattern of activations

Pat was born (1943) and brought up in Canada, studied at Pittsburgh and Oxford, and then married fellow philosopher Paul Churchland (b. 1942). They worked together at the University of Manitoba and the Institute of Advanced Study in Princeton before moving in 1984 to the University of California at San Diego where they are both professors of Philosophy, working at the boundaries of philosophy of mind and cognitive neuroscience.

Pat is known for her outspoken views on consciousness, describing the hard problem as a 'Hornswoggle problem' that will go the way of phlogiston or caloric fluid; rejecting the philosopher's zombie as the feeblest of thought-experiments, and comparing quantum coherence in microtubules to pixie dust in the synapses. She is author of *Neurophilosophy* (1986) and *Brain-wise* (2002). Paul is best known for his eliminative materialism and his rejection of such common-sense folk psychological concepts as beliefs and desires. His books include *Matter and Consciousness* (1984) and *The Engine of Reason: The Seat of the Soul* (1996).

Sue   What's the problem? Why is consciousness such a problem?

Pat   Well I don't know that it is a deeper or more difficult problem than lots of other things with regard to the brain. The fact is that we've very little by way of a fundamental understanding of the brain. Let me tell you what we don't know. We don't know how neurons code information. That's a lot not to know.

Sue I thought we knew that they code information by frequency of firing, by the closeness of synaptic connections...

Pat We don't know how the coding is done. For example, if it's frequency of firing over an interval, we don't know over what interval. It turns out that if you make the rate coding assumption, and then make your bins tinier and tinier, you find that the neuron responds to one thing at one time and another thing slightly later, so responsivity changes across time.

Sue It could be multiplexing at different intervals?

Pat Absolutely. So in some instances, what seems to carry information is latency to the first spike, in other instances it's absolute time of the first spike, but notice that so far we are only talking about coding in the axon. Tell me what is known about decoding in the dendrites. Just tell me anything that you know about decoding in the dendrites.

Sue But you have slipped from my question 'What's so special about consciousness?' to something we don't know about the brain. Now it seems to me, and to many people, that there's something special about the problem of consciousness. There's that deep red of those bougainvillea outside and it feels as though *I'm* experiencing that deep red. Understanding this seems to be a completely different problem from what seems like the potentially soluble problem of how the coding is achieved.

Pat I don't see how you can tell, by looking at a problem, how difficult it is. Many people suppose that by sheer contemplation of a problem, they can tell whether it is hard or easy. This is self-deception, and usually self-aggrandizing self-deception, to make it worse.

There are lots of examples where people were convinced that one problem was unsolvable, while some other problem was a trivial problem, and they turned out to be wrong about both. So consider, for example, the perihelion of mercury; it seemed like it was just a little nothing at all problem, right, that should just sort itself out in the fullness of time. But, of course, it took an Einsteinian revolution to solve it.

The problem of how proteins fold was thought to be an easy problem; whereas the problem of how information is copied from parent to offspring was thought to be really, really hard. Well, it turns out that the copying problem was basically solved between 1953 and 1960, but we still don't know how proteins fold.

No problem can say to you, 'I'm extremely difficult. You're gonna have to have a revolution to solve *me*.' People think that because we don't understand how consciousness is produced in brains, this must be telling us something really deep and interesting.

Sue So you think it's not. I take it you're referring to Dave Chalmers' distinction between the easy problems and the hard problem.

Pat Oh, his presumption strikes me as ridiculous. It's a very hard problem to know how information is coded in the brain. Is it harder than the problem of consciousness? Nobody can tell just by looking.

Sue But you haven't actually answered my question 'What is the problem of consciousness?' Paul?

Paul An obvious point to begin with is that we'd like to know the difference between being awake and being asleep. We can monitor the brain in various ways when we're asleep, or when we're awake, and we can see a considerable variety of differences; but why those differences should result in the subjective difference between having no consciousness at all and reflecting on Fermat's last theorem, or savouring the qualia of the bougainvillea, that doesn't pop out of the story. So we end up scratching our heads and saying 'Well, OK, we'll come back to that.'

Or, what about paying attention to something as opposed to not paying attention. Or, what about keeping something in short term memory for four or five seconds because it's important to an ongoing activity. I've already mentioned three elements of consciousness and you can probably start thinking of more. It's not clear how they knit together. It's not clear how the brain produces them.

Sue You mentioned there the 'qualia of the bougainvillea'. Can you say something about what you mean? It sounds as though you're happy to use the word 'qualia' which some people aren't.

Paul I'm happy to use the word qualia to describe, or to index, the fact that there are profound differences between my various visual sensations; sensations of green versus sensations of red, sensations of yellow versus sensations of white and so forth. There are differences in my olfactory sensations, my gustatory sensations, my tactile sensations. All of those are what make life worth living. Not only do I think they exist, I revel in them, I seek them out.

Sue But doesn't it make your head hurt to think about the problem of how that feeling of red—of what the redness is like for you—can relate to what's going on in the brain?

Paul It used to. It used to. I remember many years ago, I would look at it, and my jaw would drop. But in the 40 years that I've been in this business, I've learned a good deal of the history of science; the history of astronomy, physics, chemistry, and biology, and I discovered that the kind of intellectual befuddlement we feel now when we look at consciousness or qualia is by no means a new thing. Many people think that it is; they think this is the most unique problem in the entire universe. Wrong!

Take the problem of *light* a couple of hundred years ago. In speaking of light's Divine Creation in *Paradise Lost*, John Milton reverentially describes it as 'this ethereal quintessence of heav'n' (Book III, 713–16), and as 'pure' (IV, 150–54). Recall also the first three entries in *Genesis*—the source of so much of Milton's faith—to the effect that light was the very *first* of God's creations. In vain, then, should we try to explain light in terms of something that He created only *later*.

From this perspective, the modern scientific suggestion that light, which you can see by lifting your eyes to the sun or the moon, might be nothing other than the same obscure phenomenon—electromagnetism—that makes compass needles wobble, draws iron filings to a magnet, and makes little pieces of paper jump up to a charged comb, would seem ridiculous on its face. How could such a grand and obvious thing as light be identical with such arcane and apparently *invisible* obscurities?

Or—I give you another example—the famous philosopher, Bishop Berkeley, laughs at the idea that sound is a compression wave train occurring in the atmosphere. He appeals to the qualitative nature of sound, and pooh-poohs the compression wave theory because, after all, that's just particles moving back and forth.

Sue Do you mean that it was almost the same as the hard problem? It was almost explicitly about the *experience* of light or sound.

Paul Yes, that to which you had direct access of some kind. With light, direct access with your eyes; with sound, direct access with your ears; with the inner quality of our pain, direct access by introspection.

Sue So could I accurately paraphrase you as saying something like this ... historically a lot of problems which have been solved, such as getting rid of caloric fluid, getting rid of the *élan vital* or vital spirit, understanding light and sound, were actually very similar at the time. They all had inherent in them subjective versus objective. And they were solved, and went away, and you think the same thing will happen with the hard problem?

Paul That's almost exactly right, but the contrast wasn't so much then between subjective and objective; for light, it was 'visjective' versus objective, if you like. The deceptive idea was of this special epistemological window, vision, that alone gave you access to light, to an ontologically distinct kind of stuff. You may talk about electromagnetic fields oscillating, some will say, but that's changing the subject, you're not talking about *light*, that which we can *see*.

But, I'm sorry, it turned out to be just the other way around. It turned out that presumptively 'visjective' light was indeed electromagnetic waves. And, to return to inner qualia, it looks like the 'subjective' visual sensation of redness is going to be a particular pattern of activations across your opponent process cells in the LGN or V4. Think of it, if you like, as a musical chord struck across a population of neurons. There are keys in V4 and a particular pattern codes for red, a particular pattern codes for green, and so on. And that's what a subjective quale is.

Sue I would like to be clear how this relates to correlation, cause, and identity. There's a huge amount of work going on at the moment on the neural correlates of consciousness, and a lot of confusion about correlation, cause, and identity. Where do you stand on this?

Paul The easy way to cut through all that is, once again, to draw lessons from the history of science. Electromagnetic waves don't *cause* light; they're not *correlated* with light; they *are* light. That's what light is. Similarly with sound: a sound of middle C isn't *correlated* with a compression wave train of 263 Hz. It *is* a compression wave train with that frequency. And the feeling of warmth from a coffee cup isn't something that's *correlated* with mean molecular kinetic energy; it's *identical with* the mean molecular kinetic energy of the molecules in the cup.

Sue But you can't say that for colour! If we come back to the bougainvillea, you can't say that that colour is equivalent to so many nanometers or whatever. You need a particular sort of visual system interacting in a particular way with a particular mixture of wavelengths. Does that change the argument?

Paul No. There is a problem in the case of objective colour, and it's the problem of metamers. There are too many different patterns of power spectra that will produce in us exactly the same sensation. They all look red, but they're interestingly different. However that's a problem that can be solved, too.

We're not talking here about an objective colour out there on objects. We're talking about the sensation of red. And I'm willing to make the suggestion that this case is going to turn out to be exactly parallel to all of these other cases. To have a sensation, a visual sensation, say in a little circle right in the centre of one's visual field where the fovea is, is to have all of your three kinds of opponent processing cells showing a certain pattern of relative stimulation. They are blue versus yellow, red versus green, and black versus white, and all of them have heightened activity or lowered activity. The pattern of activation for red will be, say, 50%, 90%, 50%, across the three kinds of cells.

Pat I think the point is that in the early stages of a science you try to make correlations between likely events. When you're considering a phenomenon you use many different measuring instruments to get at it; so single cell recording is one, functional MRI is another, report by somebody is another. There are many different ways of getting at it. Then once we have a much richer and fuller understanding of the brain, not just with regard to consciousness but in all of its dimensions, then there may be a fit. We'll be able to say, as we did in the case of light or temperature, 'Aha, this is it. This pattern of activation in this context when the brain stem is doing such and such, that just *is* a sensation of red.'

Sue But now help me with this. When you say that light just *is* electromagnetism, or heat just *is* mean kinetic energy, I don't personally have a problem with that. I don't have any emotional difficulty, or inner problem with it at all. But when you say that my subjective experience of the view of the pool out there just *is* a pattern of neural activation, I do have a problem. Now, do you think that two hundred years ago the scientists had a similar difficulty?

And what do you think gets rid of that feeling from the point of view of the person thinking about the problem?

Pat But people *did* have the 'emotional difficulty' with the idea that light is EM waves. I think the emotional ease or difficulty really depends on how young you were when you learned the theory!

I actually see this in my undergraduates already, because they have grown up at a time when so much more is understood about the brain.

For them the brain is the thing that changes during addiction, or that changes during depression, or that changes during learning. When I say to them 'Guess what, in all probability it's gonna turn out ...' and then I make this identity claim, they're not particularly surprised. But you have to bear in mind that lots of people in the early stages of any scientific theory are very surprised. When people were told that the earth moves they thought this was hilarious; it was ludicrous; it was inconceivable; this is the thing which paradigmatically doesn't move.

Sue **This sounds slightly hopeless—that we've got to wait an awfully long time until people die.**

Pat We may not. We don't really know how long we'll have to wait.

Paul We learn from history that people don't have to die. You will probably find it relatively easy, compared to the subjective qualia case, to swallow the idea that Pat's voice is Pat's voice because it has a particular power spectrum. You are also prepared to agree that a certain musical chord that I might play for you on the piano is a very pretty sound. You probably won't appreciate that it's four different notes; that a C7th chord is a C, an E, a G, and a B flat struck simultaneously. You might initially be surprised to learn that those beautiful sounds are made up of discrete elements; that a C7th chord is one *four*some; an A minor chord is another *four*some, and so on. These sorts of appreciations are something you initially apprehended in an inarticulate way. You learn to recognize Pat's voice but you have no idea how you recognize it; you learn to recognize two different musical chords but you have no idea how you discriminate between them. Then you discover that they do have internal structure and other parts of the brain are sensitive to that internal structure and that's how you manage to discriminate them.

Pat ... and that must be true of colours too, because you've just got three cones and the opponent process cells. So, when I look at yellow, I may think 'yellow is just yellow', but in actual fact it is a kind of composite. It really is.

Paul It's an activation vector across three different kinds of cells.

Sue **What about pain?**

Pat In the beginning, people said that there's the *sensation* of pain and the *awfulness* of pain, and they can't be dissociated. I knew philosophers who said that it was a necessary truth that pain was awful, and pain was awful in all possible worlds. Now people just routinely accept that pain is dissociable in those ways, even though normally it doesn't seem that way.

Paul It's called codeine.

Sue Or heroin.

Paul And it does make you no longer give a damn.

Sue Yeah, but why do I not have much of a problem with some of those examples? Well actually, no. Why do I have no problem at all with the auditory example, a bit of a creeping bothersome problem with the colour example, and a really, really big problem about how neurons firing in the anterior cingulate cortex can be this awfulness of pain?

Paul Because you're climbing a knowledge gradient.

Pat And you're way down the hill.

Paul If you knew enough about the brain, and how it codes, and how the space of possible coding vectors maps onto the space of possible colours, and the space of possible coding vectors in your pain registering system maps onto the space of possible nociceptive stimulations, then you start to see that the activity in these various parts of the brain is in fact a highly sophisticated map of the external feature space. You start to get a grip on how it's a representation, and it no longer seems quite so mysterious. In advance of that, of course, you're just left clawing at the air.

Sue Does this creeping up the knowledge gradient lead in the direction of doing away with dualism? Because I often feel that I'm falling again and again into some kind of dualism between inner and outer, or subjective and objective, or me in here and the world out there.

Pat There is a real dualism here, but not one involving spooky stuff. One of the things that your brain does is build a model, and within that model it marks the difference between what's inner and what's outer. In my brain's distinction between inner and outer I always have an efference copy of a command to make a movement. So I always know the movement is mine, and I can't tickle myself. But schizophrenics can. Something is wrong with their system for efference copy.

Paul They don't know where their self leaves off and the independent world begins.

Sue I want to change tack completely now. One at a time, do you think that a philosopher's zombie is possible?

Pat Well if you mean, is it...

Paul Say no.

Sue Now you're not to ... As close as your views might be, you're not to tell each other what to say.

Pat It depends on what you mean by possible. Of course it's logically possible, but that's not interesting. We're not really interested in whether somebody can write a story about somebody who's a zombie; we're interested in knowing whether or not it's empirically possible. And it does not seem to be, so far as we know. People in coma, or deep sleep, or absent seizures, do not have awareness. And the behaviour in those three conditions is very different from the behaviour when people are awake. Now it could turn out that there is somebody who is a zombie, but that's like asking 'Could it turn out that there's a whole species of animals, where none of them have DNA?' Logically that's possible, but from everything we know about natural selection it's just not likely.

Having said that, I'm also in great admiration of the work that Mel Goodale and David Milner do, which shows that some part of the motor system can use nonconscious visual information. Christof and Francis (in my view unfortunately) called the system that Goodale and Milner study a 'zombie system'.

Sue But Goodale himself doesn't call it a zombie system. That seems to me the whole point of the distinction they're making; that it's action versus perception; not conscious versus unconscious, and that seems to me a great step forward.

Pat Exactly. That's why I said that Christof and Francis have *unfortunately* called it that. I think the work is brilliant and is some of the most interesting work on consciousness that there is.

Sue Paul, you said 'Just say no' ... to use a popular American phrase. Could you explain?

Paul Sure. Once again, here's a parallel: someone could say, 'Look, light can't be identical with electromagnetic waves because I can imagine a universe in which electromagnetic waves are bouncing about all over the place, but it's pitch black from one end to the other.' It's a zombie universe if you like, only here it's light that's missing.

And one wants to say, 'Well you can imagine that all you like, but the question here is, what is light as a matter of fact?' And the truth is that when you learn about light, and about electromagnetic waves, and how they make plants grow and make sunflowers point towards the stars, it turns out that this universe, which is supposedly devoid of light, behaves exactly like the one we're in. Everything in it behaves as if the stars are shining like mad, thank you very much. So the more you know about both light and electromagnetic waves, the harder it is to coherently imagine a universe that is abuzz with electromagnetic waves but is dark.

Similarly, the more we learn to understand how the brain works at a low level, and the more we learn to understand the psychology at a high level, the more we'll see how they fit together in this wonderful embrace such that they're not two things embracing one another, they're actually just one thing, looked at from two different points of view. So the more we learn about the brain, the harder it will be to enjoy Chalmers' thought experiment.

Sue **So do you think that, because brain research is going so fast, fairly soon people just won't fall for the zombic hunch?**

Pat Oh sure.

Paul Exactly right. It's an argument with an illegitimate appeal, and the illegitimate appeal derives from people's ignorance. So as the ignorance slowly fades, the appeal of the argument will fade.

Pat The other thing that I personally find unappealing about Chalmers' view, is that I'm not terribly excited when philosophers tell me there's something science can't do. Colin McGinn and Jerry Fodor, and a lot of these guys, are making a living by pronouncing on what science can't possibly do, and David Chalmers made his name by saying you'll never explain consciousness neurobiologically. I think it's a defeatist way of going about things, but it may appeal to philosophers who are afraid that neuroscience is taking over their business. It's much more interesting to try to come up with a positive theory of something, and what I have not yet seen from Chalmers is any sort of positive theory that explains qualia. If he thinks it's a fundamental feature of the universe along with mass and charge, as he sometimes says, then let's go after that and let's do the science.

Sue **You mentioned qualia again, and I detect here a difference between you and Dan Dennett. Lots of people assume that because all three of**

you are some sort of materialists, and have ideas in common, that your views are identical. Could you briefly explain where you differ from Dennett in your views?

Pat   Well I'll say a little bit and then let Paul amplify.

Dan really does have a very different perspective. From his perspective, the perspective of behaviourism, you don't need qualia in the story. All you really need is reportability. And so you have a conscious phenomenon if and only if it's reportable. I look at it from a much more biological perspective, and it seems to me that there really are these qualitative experiences, and many of them are actually generated internally. Such feelings as hunger and thirst, and lust and curiosity, are not stimulus bound in any way, and you want to be able to tell the story about those. So I think that there are various qualitative experiences and there are brain states of some kind to which they are identical. And the problem will be to have a sufficiently rich theory in neuroscience so that we can specify to which neurobiological activity a state of *lust* is identical. Or to which activity being *fatigued* is identical.

Sue   But I still don't understand the difference between you and Dennett, because Dennett would not be a behaviourist in the sense of saying that all you need is behavioural report. He would say that it's also legitimate to do brain science and study what's going on in the brain. I think he would say that all of that is sufficient, and when all that's done we'll realise that there never were any qualia—in the sense of some sort of separate 'ineffable suchness' of the red. I don't see where you're different.

Paul   Oh, we would agree with him on that. The way philosophers have characterized qualia, they've mis-characterized them: they're 'known incorrigibly', they're 'ontological simples'. When Dan says nothing like that exists, I'm inclined to agree with him, but I think what he's eliminating is a philosophers' creation. What's real in you and me and anybody who looks at the red we were looking at, is some sort of activational state of your visual system. That's entirely real, thank you very much.

Pat   Dan's absolutely right about this, if you define qualia as non-material, ineffable essences, then gee shucks, that is pretty darn mysterious. But at other times he sounds like he really is a behaviourist. So in all honesty I think I'd have to say that I'm not always sure exactly what he means.

Sue Do you think consciousness survives the death of the physical body?

Pat We do know that when large numbers of neurons die, as in Alzheimer's disease, deficits in memory occur, cognition is impaired, personality changes, awareness of what other people are thinking and feeling, and awareness of time and place, are impaired. I see this as a kind of fading of many aspects of the self and its capacities, and one cannot but feel that the person one knew and loved is no longer there. All the evidence shows that the brain is necessary for functions associated with consciousness. I am not sure how consciousness could survive the death of the brain if it needs neurons to sustain it.

At a personal level, I should say that I feel more settled about death and dying having understood that it is the end, than I would if I were trying to nourish an unrealistic hope in some kind of heaven. When I was a child, a friend who was a native Indian once remarked to me that he felt sorry for Christians, as they labour under the delusion of a heaven, while he, in contrast, could prepare for finality, pass on the stories of the person's life, help them to die easily, and accept the finality for what it is. That struck me as sensible then, and it does so still.

Paul I agree. Consciousness is just one particularly sophisticated dimension of biological life. When my biological life ends, so does my consciousness. I am more than content with this. The prospect of being conscious for an unending eternity is quite frankly appalling.

When my time comes, let me sleep.

Sue Do you think you have free will?

Pat If you mean 'Are my decisions not caused?' surely not. From everything we know, the brain is a causal machine. It goes from state to state as a result of antecedent conditions, and if the antecedent conditions were different, the state would have been different. But, having said that, as humans we're still really interested in the difference between behaviour that you might say is in-control behaviour and behaviour that is not in-control behaviour, and I believe that, at least to a first approximation, we can give a neurobiological characterization of the differences. We can begin to identify the relevant parameters, and we can conceive of the problem in terms of a parameter space. You can think of it visually as a three-dimensional parameter space, but it's going to be an n-dimensional parameter space.

Paul With a large n.

Pat   And there'll be a volume in there within which people are in control. It's going to have fuzzy boundaries, be a funny shape, have dynamic properties, and there'll be different ways of being in control. So that the appearance of hormones at adolescence, for example, is going to change the shape of somebody's in-control space.

Sue   But what about in your own life? I mean, that's not the problem that causes people to worry when they say, 'Well if the brain is causally closed then it doesn't matter what I decide.' This is what seems to make life, and making moral choices and decisions, difficult. How, in your life, does your philosophy relate to your actual decision making, or the things you do, or the way you feel?

Pat   I think you just hold those two things in your mind at the same time. I mean the way the brain works is that, amongst other things, it has this user illusion—that your decisions are made according to, shall we say, the standard model—that you consciously identify the options, you consciously do an expected utility calculation, you consciously choose, and then at some point later in time, the action's executed. That's a useful user illusion.

Sue   So do you mean that you're happy to think this is an illusion and then just behave as though it's real?

Pat   It's like the illusion with morality. We know that moral laws are not specified by the gods. We know that they are first of all neurobiologically based or evolutionarily based, and secondly culturally based, but it's also very useful for people to have the illusion that these are really true. Now that's a slightly different problem, but I don't have any particular difficulty in my own life in making decisions and being responsible for them. Whether it makes me happy is not the point; whether it is true, is.

Sue   What about you Paul? How do you live with this in your life?

Paul   I don't feel the conflict at all because, when you put the question to Pat, it was as if one's body is behaving and one's decisions make no difference to what happens. But that isn't how I experience my life. Whether or not my hands go up is a function of the conversation I'm conducting. My behaviour is quite regularly a product of my will. The question is 'What's behind the will? Is that being systematically caused?' I'm inclined to say 'yes', but the following thought is relevant, and it comforts me to some degree.

We know that brains are non-linear dynamical systems. These are systems that are governed by continuum mathematics, and their behaviour is exquisitely sensitive to infinitesimally small differences, such that two brains in almost exactly the same state will quickly wind off in very, very different states. This means that the brain of a human, or even of a mouse, is a system whose behaviour is unpredictable by any machine constructible in this universe. We are importantly unpredictable save for general tendencies and patterns. We will go to sleep at night, get up in the morning, tend to hug our wife at least three or four times a day, but exactly when, or what words will come out of my mouth, that's unpredictable. So one mustn't fear the story science seems to tell, that we are just robots.

Sue But you did say it comforts you. You wouldn't have said that unless there was something that you find slightly uncomfortable, and that makes you need comforting?

Paul Sure. I am just like everybody else. I would be upset to learn that, say, I'm a completely programmed robot. It's conceivable. There have been stories written about this, like Philip Dick's *Do Androids Dream of Electric Sheep?* back in the 1950s, from which the movie *Blade Runner* was made. This fellow discovered that he was in fact a robot, and was somewhat distressed by it.

Sue And you would be distressed too?

Paul If I learned I was a predictively programmed machine, yes.

Sue So, for you, the unpredictability is comforting, even though it's deterministic?

Paul Yes, because it rules out one ugly possibility; that I am someone's puppet.

Sue Pat, do you think that studying consciousness all this time has changed you as a person?

Pat You know, I don't think of myself as really studying *consciousness* all this time. I mean, my interests are really diverse, and a lot of them are squarely within neurobiology.

Sue Then take the question more broadly—how has studying all those things changed you as a person and affected your life?

Pat Well it's hard to say introspectively but, like everybody else, I think that the developments in neuroscience have had a big impact on how we think about all kinds of things, especially pathological cases. When I was a kid, people used to think that autism was the result of cold mothering, people talked about nervous breakdowns, and even when I was an undergraduate people thought that depression was something that you should be able to cure through Freudian analysis. So there've been enormous changes, and in my experience with run-of-the-mill people—you know, people who cut my hair, or the people I meet on dog beach—they're all interested in the brain. Everybody has a brain, and everybody has somebody in their family with pathology of some kind or other. And I just find almost any aspect of neurobiology endlessly fascinating, whether it's the medicinal leech, or the rhesus monkey, or the human.

So in one sense, of course, it's changed my life profoundly, but in other ways it hasn't. I mean I still love my family, and I'm just monumentally excited when the grand-babies are born. I still like dogs. I'd still rather be canoeing than going to a museum, and living in the bush is still for me, the greatest thing that one can do.

Sue Ah, you're still human after all!

Pat Another interest of mine that's connected to the free will problem is how the developing knowledge of the genetic and neurobiological causes of irrational violence is going to have an impact on the criminal law. For example, there are MAOA mutants who, if they have an abusive upbringing, are virtually certain to be irrationally and self-destructively violent. So there are some very difficult questions about how we best deal with them, especially if it turns out we can intervene. The interventions may not always be pretty, but of course going to prison is not pretty either, especially in America. Also drugs to deal with addiction are just around the corner, and that may mean that we have very different possibilities for changing the drug laws than we do now.

Sue What would you like to happen to the drug laws?

Pat I think that the drug laws are really self-defeating at this point. They generate a huge underworld of criminals and they don't prevent people from taking drugs. So what you really want to do is collect taxes from the sale of drugs, have standards so that people get purer stuff, if they must have it, and then educate them as well as you possibly can to tell them about the dangers. You know now we can't go into a ladies' loo without seeing a sign that says 'Don't drink if you're pregnant'. Well that's terrific, but we could also have a sign that says 'Don't take cocaine and don't shoot speed if you're pregnant'. Why not? So I would love to see the drug laws change. It would mean that the prison population would be cut in half.

You see that's a very practical thing, but practical changes in the criminal law come about as a result of scientific changes. To take a slightly different case, views on homosexuality have changed enormously, with the understanding that it's not, as we used to say, 'a lifestyle choice'. That's just garbage; people's brains are as they are. So people's attitudes, especially in the younger generations, are completely different from those of my generation. I suspect something similar will happen with drugs, and in Canada we can already see those changes in legislation.

Sue And Paul, how did you first get interested in consciousness?

Paul I was an undergraduate in physics and math, and sort of discovered philosophy along the way. It slowly came over me, and I was captured by issues in the philosophy of science and epistemology: how does the human race learn as the centuries roll by, and how is this knowledge embodied? For the logical positivists, their paradigm of representation was language—forgetting that we're the only creatures on the planet that use language, and that even humans represent the world without language for their first two years.

So slowly I begin to appreciate, because I was a naturalist, that it's the brain that's ultimately doing all this. And brains talking to one another, and brains building a culture that they leave behind, led me to be interested in the brain. Pat got interested in the brain big time back around 1975, and ran way ahead of me there for a while, and then we found ourselves looking more and more at the empirical data, and at the theories. As they have grown over time, I think it's had a profound effect on my philosophical views. The epistemology I now defend is radically different from the one I would have defended as a very young man.

Sue But what about yourself, in your own life?

Paul Sure. Well for one thing my views on moral knowledge have changed profoundly. I now am a robust moral realist, and I regard that one of the things the brain learns, perhaps the most important thing any brain ever learns, is how to perceive other minds and other people.

One learns to navigate social space, as well as navigate physical space; one learns to find a nest in social space just as one finds a nest in physical space. One acquires the skills of moral perception and social perception. Other people have different offices, stations, and obligations within moral space, and sometimes they're the same as yours; sometimes different. So you find hopelessly out-of-control scoundrels, and well-controlled people you can trust in an emergency, people with whom you can make valuable community. You've got to keep track of all of that. These are skills that brains learn to have. Understanding how the brain works is understanding how all that happens.

Sue And how does that change your own life and the way you live it?

Paul It gives me a different perspective on the occasional conflicts I run into. It gives me a different perspective on the variety of personalities I encounter. It makes the successes that surround me, like my wife and my children and my friends, far more precious to me than they would otherwise be, because I appreciate how difficult it is for brains to succeed in these many endeavours, and how much one's own success depends upon luck.

I'd like to know something that I don't understand, and will hopefully learn more about before we die: What happens to married couples over long periods of time? What is the nature of the very special community that's made there? I sometimes wax romantic on this. I'm given to be a bit of a romantic in any case, and I am willing to be more romantic in any language, including the new, and allegedly austere, language of neuroscience. I don't think it's austere at all. I think it holds the promise of giving us moral insights that we could not get in any other way.

Sue I can see your interest in marriage, because you are a rather unique couple aren't you; being so close philosophically as well as married?

Paul We are for our generation perhaps. In fact married couples in academia were prohibited at the University of Toronto where I had my first job. I would still be in Toronto but for their nepotism rules! But now married couples, either working very closely together or in various complementary ways, are common. In fact it's often a really good opportunity, you can get two for one.

Pat It has been fun actually; it's been enormously good fun.

Sue And still is, by the look of it! I don't like the past tense there!

Paul As the neuroscience gets better it's also affected the way I look at other things that I've loved. I was a musician as a young man, and when nobody's at home I'll still sit down and play the guitar, theorizing about the cognitive neurobiology of music, of music appreciation, of music composition, or simply the skills of playing an instrument, worrying about how the brain does these things.

Sue So do you mean the music is enriched by knowing about the brain, rather than diminished?

Paul Oh yes, yes. People are inclined to think it must be diminished, but again it's the knowledge gradient that they're climbing. If they're asked to conceive of something they love, say opera, in terms of some brain theory of which they have essentially no comprehension, they immediately think 'Oh that must be eviscerating'. But no, dear, it's just the reverse.

Sue So you're with Richard Dawkins on his *Unweaving the Rainbow*?

Pat Absolutely. People get their spirituality (in the secular sense) in many different ways. We get some—only some—of ours through science.

Paul Yes. The world is a much richer place than we think it is, and the only way we can discover that richness is to learn more about how it works.

Francis Crick

You're just a pack of neurons

Sir Francis Crick (1916–2004) is best known for his collaboration with James D. Watson in their discovery of the structure of DNA: the double helix. They received the Nobel Prize in Medicine and Physiology for this world-changing discovery in 1962. Originally studying physics in London, Crick spent the war years working for the Admiralty. He left in 1947, wanting to pursue the mystery of life and the boundary between living and non-living things, and so trained in biology, getting a PhD in X-ray diffraction at the University of Cambridge in 1954. Years later he changed tack again and began theoretical work on vision, the function of dreams and the nature of consciousness. Until his death in 2004 he was a professor at the Salk Institute in La Jolla, California, where he collaborated closely with Christof Koch on their search for the neural correlates of visual consciousness. He is author of *The Astonishing Hypothesis* (1994).

Sue Why is consciousness such a problem?

Francis There's no easy way of explaining consciousness in terms of known science. The easiest way to talk about the problem is in terms of qualia. For example—how can you explain the redness of red in terms of physics and chemistry?

Sue It's interesting that you begin with qualia because some people, such as Dan Dennett, think the problem is so serious that we have to get rid

of qualia altogether, while other people claim that we must solve the hard problem and explain how qualia are generated by the brain. How do you personally think about the problem of qualia?

Francis The line that Christof and I take is that we shouldn't approach the hard problem head on. We should try and find the neural correlates that correspond to what we're conscious of. Let me just say that much of what goes on in our brain is unconscious, and therefore what we want to know is the difference in the activity of the brain when you're conscious compared to when you're not conscious. Philosophers may think they can explain it, but all they do is argue about it without actually finding out what's going on.

Sue Yet there seems to be a really difficult problem here doesn't there? You just talked about the difference between conscious and unconscious processes. So let's imagine our own brains as we sit here in your house. There are multiple parallel processes going on inside our brains and some of those give rise to, or are correlated with, our experience of seeing the blue of that pool out there, while others aren't. Doesn't this seem to be a completely magical difference; an insoluble problem?

Francis Well of course that's what people say. That's what they said about life. They said there was a vital spirit that you couldn't explain in terms of physics or chemistry, and because they said it, it became almost a standard point of view.

Sue So do you think consciousness is a similar problem?

Francis It's an analogy, and the history of the vital spirit, or the *élan vital*, shows us that you have to be cautious.

Sue So you're setting aside some of these difficulties and saying that the best way forward is to measure the *correlates* of consciousness.

Francis Yes, but we must be clear about the word 'correlate'. If A is correlated to B, then B is correlated to A, in other words it's reciprocal. The question is whether it's causal—whether the cock crow *causes* the sun to rise, or whether it's *correlated* with the rising of the sun or, more sensibly, the other way round, whether the sunrise causes the cock to crow.

So, strictly speaking, we're interested in the causal stage. But in the first instance you look for the correlates and then you go on to look at cause. That's standard, that's actually what scientists do. That's what they call a controlled experiment.

Sue **So let's take an example of an experiment on neural correlates that has been done.**

Francis Binocular rivalry is the standard.

Sue **Right. In experiments on binocular rivalry you find that if one percept is dominant a certain group of cells is firing, and with another percept dominant another group of cells is firing. What would you want to say about that correlation? And how would you move from correlation to cause?**

Binocular rivalry. When different images are shown to the two eyes they usually do not merge into one but compete for dominance. In this case the experience alternates between seeing horizontal and vertical stripes. In the 1980s Nikos Logothetis and his colleagues performed experiments in which monkeys pressed a lever to say which image they were seeing. They then recorded from single cells in the brain and showed that in the early parts of the visual system nothing changed when the experience changed, but further up the visual hierarchy different cells were active depending on which image the monkey reported seeing. Similar results have subsequently been found in humans using brain imaging. But what does this mean? Are these areas the seat of consciousness or the place where consciousness happens or is generated? Or does this way of thinking about it imply a Cartesian theatre?

Francis First you want an idea of whether it's that set of cells firing, or whether they fire in a special way, or whether it's a combination of the two, or something else quite different. In other words you've got to have a working hypothesis.

We have what we call a framework, or set of working hypotheses, in which we think we see the general shape of what's happening. Otherwise you don't know where to begin. To take the case you mentioned, we might ask—is it where it is that matters? Is it firing at a particular frequency? Or is it something more complicated than that?

Sue And using that framework, how would you move from studying correlations to understanding causes?

Francis That's exactly what we hope we're going to do. Let's say what it involves. It involves putting many electrodes in all the areas in the hierarchy and seeing the way the interactions change with time. Then you can try to see a causal interaction between something in one area and something in another area, since a cause must come first. You see it goes up to some higher level of the visual hierarchy but it's not quite clear whether it has to go to the front or the back of the brain first. It depends on what you're looking at; what you're concentrating on, or are interested in—then that's the first thing that becomes visible.

Take an example, if you flash a quick look at this room, people will say 'That's a room with a piano, and two chairs.' They'll give you a general impression, but they won't be able to tell you how big the table is, or details of that sort. So the idea is that the first thing that becomes conscious is at some higher level and that then signals back, and you gradually become conscious in more and more detail.

So, in other words, you travel up the hierarchy unconsciously, and then you travel back down it again consciously.

Sue Over what kind of time course are you talking about?

Francis 100 milliseconds.

Sue Does this relate to other timings such as Libet's half second?

Francis Probably.

Sue Could we return now to your framework of hypotheses and your question of whether consciousness depends on which neurons are firing, whether they fire in a special way, or perhaps something quite different. Could you just briefly outline to me how you think the evidence is going on that question?

Francis I think there's a general consensus that it's due to the correlation of some coalition of neurons; that you have to form a coalition of neurons. People call it various things; Edelman calls it the dynamic core, and others have the same basic idea.

We work mainly on the visual system, and in vision there are usually a number of alternative interpretations of the visual input. The brain has to decide which of these alternative interpretations is the most plausible, and that's what it's going to see—in fact that's probably what it is going to act on, whether it sees it or not.

Sue **But here we get to the crux of what feels such a difficult problem in consciousness. It seems easy to understand how coalitions might form, and how the decision that one model is better than the other might lead to action and interaction with the world. Yet intuitively it seems that subjective experience is something altogether different.**

Francis Yes it probably is, but you have first to understand what's going on, which is the question you've asked.

Sue **So are you, as it were, leaving subjective experience on one side and getting on with the job of finding out how the brain works, in the hope that the problem will just one day be solved? How do you really feel about the hard problem of subjective experience?**

Francis We believe that, at any moment, there's a coalition of neurons which are firing together; firing more or less at the same time, and probably above some threshold. When you're seeing a particular scene, as Edelman and Tononi are fond of pointing out, that is only one of an immense number of scenes you might possibly see.

For example, I might now be thinking about motorcars, but at the moment before I said that I wasn't. You see, there's all that activity which isn't going on. Therefore the NCCs depend on a minority of neurons at any one moment. It's a subset: a relatively small number of neurons. We wouldn't like to say what percentage it is, but one percent, ten percent, or, some numbers like that.

Sue **But that's still a lot of neurons.**

Francis Yes, but it could be less. The point is, I want a small number of neurons that is firing away at any particular moment and corresponds to the NCC. But how many neurons is that connected to?

Sue **Well in some way the whole brain presumably.**

Francis Well, at least, shall we say a thousand times as many. No I don't think it is the whole brain for the same reason—not directly at any rate, because of all my many associations. For example, there is the motor car I mentioned which is associated with sitting here and talking to you, you see. So that's what we call the penumbra and by definition that's unconscious.

So, in other words, one of the things which you arrive at with these models is that because of the nature of the brain, and the fact that one neuron connects to so many others, this must mean that there's a large number of neurons which have been associated. They can become conscious if the NCC shifts. So the penumbra is the unique feature of the brain as compared to today's computers.

Therefore we say that if you're going to attempt the hard problem, you've got to consider the nature of the penumbra. Well that gets us one step in that direction, and it shows you that we have made a step which we hadn't made before.

Sue **Can you tell me how this penumbra relates to Global Workspace Theory, where you have a bright spot on the stage of the mind's theatre, and a shadowy fringe around the bright spot?**

Francis I think that it's all rather vague, and the present ideas are much more precise. If you look at our framework of ideas, you can see that the global workspace is the idea behind them, but they go much further than that.

So I think Global Workspace Theory started people thinking about lots of things interacting together, and now we've got to the stage where we can ask questions—how, and which way do they interact? And we can carry out experiments to show the dynamics of the interaction.

For example, we know that in binocular rivalry, if the stimulus is changing in appearance, there's a wave of activity travelling over the cortex which corresponds to that change. You can't arrange the brain so that everything is simultaneously in high speed contact with everything else, so there must be time delays. We're talking about 50–100 milliseconds—that sort of times. So then we can ask—can you see those changes? In other words, once you have a postulate of this nature, you can do experiments to show what's actually happening. And the way you ask the question is not the way philosophers normally approach the problem.

Sue **Are you encouraged by the speed with which all this is going, and the findings so far?**

Francis I do think what has happened in the last year or two has been encouraging. We've been reserved for so long that I'm reluctant to say that, but yes, I think it has been good.

Sue You mentioned the role of philosophers, what part do you think philosophers are playing, or have played in this?

Francis Well we have an established series of jokes about philosophers which I don't have to give! Essentially philosophers often ask good questions, but they have no techniques for getting the answers. Therefore you should not pay too much attention to their discussions. And we can ask what progress they have made. A lot of problems which were once regarded as philosophical, such as what is an atom, are now regarded as part of physics. Some people have argued that the main purpose of a philosopher is to deal with the unsolved problems, but the problems eventually get solved, and they get solved in a scientific way. If you ask how many cases in the past has a philosopher been successful at solving a problem, as far as we can say there are no such cases.

Essentially, their main technique is the thought experiment, and here you can argue indefinitely. Let me give you an example—John Searle's Chinese room. You see I think this shows just the same disadvantages. It says that if you have a system that can only deal with syntax, it can't deal with semantics. Once you've said that, you've said it all, and you haven't proved it anyway, you see.

An exception is two cases that were done by a man who wasn't normally thought of as a philosopher, and he didn't think in terms of words as philosophers do, but in terms of equations and visual images … and that was Einstein.

Sue So there's a big difference in your mind between Einstein's thought experiment of sitting on a light wave, which led directly through mathematics to a new vision of the world, and a thought experiment like, let's say the philosopher's zombie. Do you believe that a philosopher's zombie is possible?

Francis No, I don't think it is. Because we've got clear ideas now of the sort of thing that's needed to be conscious. You have to be aware of something for a certain limited period of time, and with a situation sufficiently complicated you need to have the opportunity of responding to it, or dealing with it, or thinking about it in a number of very different ways. Now a zombie system, in our terminology, is something which is much more stereotyped and automatic. We believe there are such modes in the human brain; for example when you're sleepwalking is one case, or when you nodded your head just then, that was a zombie response. So we use it in that sense.

But if you abolished consciousness and asked what would a person actually be like—they would be like sleepwalking.

Sue **So you couldn't have a person who was behaving normally, and their brain was doing the things that you're talking about and yet, somehow, they weren't conscious.**

Francis No. Contradiction in terms. I wouldn't spend any time on it.

Sue **Can I turn now to some more personal questions? How did you first get interested in the problems of consciousness?**

Francis Well, that's a long complicated story. My career was interrupted during the war because of doing war work for the British Admiralty, and after that I had to decide what to do. I decided eventually that I didn't want to go on working for the Admiralty, producing weapons and things of that sort. But I accepted a job as a permanent civil servant after the war. So I had a job, but it wasn't what I wanted to do. So then I had to make my mind up—what did I want to do?

I decided there were several problems, but there were two in particular which most people thought were difficult, if not impossible, to understand scientifically. One was the borderline between the living and the non-living, and the other was how the brain works—and that would include the consciousness aspect of it. I decided that if I was going to do something interesting, I should choose one of those.

Sue **How wonderful to have spent a life looking at those two marvellous transitions, one between the living and the not-living, and the other between the conscious and the not-conscious.**

Francis Well it wasn't actually a simple decision. It was even more odd than that. After some few weeks thinking about it, and I'd boiled it down to these two, I had to decide which one of those two. I decided that my background was really much more relevant to the living and non-living thing, as opposed to the brain, and I should really look for something in that. Well, a week or so later I was offered a job to work on the eye, and I actually turned it down on the grounds of what I'd decided previously. I think, looking back, it was just as well, and so then I applied to the Medical Research Council and the rest you know.

Sue **So at what point, having done all that wonderful work on the problem of life, did you make a decision to turn to consciousness?**

Francis Well without going into it, there were complicated reasons why I was tempted to come back here to the Salk Institute. And I decided that if I was going to change fields, this was the time to do it. By then I was 60 you see.

It took me two or three years to get away and to tidy up things, and then I chose the visual system simply out of ignorance really. But there were good arguments for using the visual system. We are very visual animals. The cat and the macaque are very visual animals, and there is a lot of work done on vision, both in neuroanatomy and behaviour. It's a minute amount compared with what we require but still, these are good reasons.

It was only later that I gradually got drawn into consciousness. You see, the experiments by Hubel and Wiesel, or Semir Zeki, were done on animals, but they were anaesthetized, they weren't really seeing anything.

Sue   Do you mean that you felt there was something important left out?

Francis   Yes.

Sue   You once wrote about the 'Astonishing Hypothesis', and the idea that 'You're just a pack of neurons'. Do you think most people still find that idea astonishing?

Francis   Most of the people who found *The Astonishing Hypothesis* to be astonishing—which is most of the people in the world and very many people in the USA—would still find it astonishing. The big change is that an increasing number of scientists now consider it, as we do, to be a genuine scientific problem.

Sue   Do you believe you have free will?

Francis   Daniel Wegner has made a good case that you're not conscious of much of what goes on—that in some sense it's an epiphenomenon. I think that's right, and I think his explanation is right too. It's a useful phenomenon. Even though it doesn't tell you exactly what's happening every time, it gives you some sort of record of the way things happened. Dan Dennett has written this long book and rambled about it, but I think Wegner's is much more to the point.

Sue   If you think that, how does that affect your life and your own decisions? For example, you might look back over all those decisions you've just told me about, concerning your scientific choices and so on. If you take Wegner's point of view, you would have to say they were made by underlying mechanical deterministic processes, and the feeling of will is an illusion. Are you happy to look on your life like that?

Francis   That's right. I think it must be deterministic. It's just that people confronted with this have chosen the wrong explanation—that there's some sort of soul or other which is separate from the brain. They're dualists essentially.

Sue   And you're a straight down the line monist are you?

Francis   Yes.

Sue   What do you believe happens to consciousness after death?

Francis   Personally I believe that it is highly unlikely that there is consciousness after death, but that, after all, is what we are attempting to prove, in as far as one can prove anything scientifically.

Sue   There's been a lot of progress in understanding the brain in the past few decades. Can you say how the way your own understanding of the brain has developed affects the way you live your life?

Francis   Well I don't think it makes much difference frankly, but I can understand why you ask these questions, because you're interested in Buddhism.

Sue   No, I don't think that's right—well it might be. Can you explain what you mean?

Francis   I think you're really trying to look for general explanations along these lines, and Buddhism is the way you want to go. You don't think so much in terms of neuroscience. I think your enthusiasm for Buddhism was there anyway.

Sue   What I like about Buddhism is that some of its central tenets seem to fit so well with what we are learning about the brain, but in addition it gives you a way of practising—and that helps you see better into the nature of consciousness. Have you ever meditated or been tempted to try any practice of that kind?

Francis   Not really, no. But the real question is what experiments do you suggest?

Sue   Well one idea that I've been playing with for experiments is as follows. If Dan Dennett's multiple drafts theory is right, then there is no fact of the matter about which of the multiple things going on in the brain is conscious and which isn't. So...

Francis   Now let me say why I think all that's nonsense; because essentially it's purely psychology and you're not talking about neurons. It must be an experiment that deals with neurons from our point of view.

Sue **So do you think that the only legitimate experimental way forward is on neurons, and that psychology can't provide useful experiments?**

Francis No, but Dennett is mistaken because he isn't using a combination of the two. Therefore, if you're basing your work on Dennett's ideas, you'll be liable to be criticized because Dennett simply isn't paying attention to neurons.

And let me say that he agrees with this—he has said that neurons are not his department. So our view is that if you won't explain it in terms of neurons it's like saying that you're interested in evolution but genes are not your department.

It's important to have the psychological stuff as well, but that's another level of explanation, and both levels of explanation have got to be right.

Sue **And would you go for the lowest possible level of explanation? Is that the sort of explanation that would make you most happy?**

Francis Oh yes. Eventually you've got to get down to neurotransmitters and things like that, you see. And it's a nice question whether consciousness is due to the concentration of calcium in a particular type of cell. That's not the whole explanation, but it's part of the explanation, and it may be a crucial part.

Sue **If you were to be here in 50 years time, what would you like to see has been achieved?**

Francis I would want to see where the field had got to, but you can't see that in advance. Suppose you'd asked that question in 1918, for example. At that time, one of the leading geneticists in England said that you'll never explain genes in terms of chemistry.

Sue **Yes, but by saying that, you're implying that he would have liked to see genes explained in terms of chemistry, or somebody would have liked to see it. So what's the equivalent now to that statement?**

Francis We would simply like to see what the cause of consciousness is. We'd like to have a description of it in scientific terms, but what the description is you can't tell in advance. I remember someone being asked at his inaugural lecture—what is the next important step? And he said, 'Well if I knew that, I'd take it.'

Daniel Dennett

*You have to give
up your intuitions
about consciousness*

Dan Dennett was born (1942) in Boston, studied at Harvard, and took his DPhil in Oxford in 1965, where he studied with Gilbert Ryle. Since 1971 he has been at Tufts University in Massachusetts where he is Director of the Center for Cognitive Studies. In the field of consciousness studies he is best known for his rejection of the Cartesian theatre in favour of his theory of multiple drafts, and for the method of heterophenomenology, but he has a long standing interest in artificial intelligence and robots, in evolutionary theory and memetics, and in the problems of free will. He spends the summers on his farm in Maine where he thinks about consciousness while sailing, mowing the hay, or making his own cider. Among his many books are *The Intentional Stance* (1987), *Consciousness Explained* (1991), *Darwin's Dangerous Idea* (1996), and *Freedom Evolves* (2003).

Sue Why do you think consciousness seems to be a harder problem than many other problems in science? For you, what's special about the problem of consciousness?

Dan Human brains are just the most complicated thing that's yet evolved, and we're trying to understand them using our brains. There are people who have suggested that this was impossible. That's just nonsense, but I think the reason that we find consciousness so hard is that we have evolved a certain capacity for self-knowledge, a certain access to ourselves which gives us subjective experience—which gives us a way of looking out at the world from where we are. And this just turns out to be very hard to understand.

How can something have that perspective? It might be just a thing, but it's a thing with a point of view, and with the capacity to reflect on that point of view and talk about it. Each one of us is trapped within a point of view. I can't ever get inside your head, and you can't ever get inside mine. The undeniable fact that we have these perspectives is not closely parallelled with anything else we know about anything else. It isn't that atoms have that sort of thing, or that molecules do, or that volcanoes or continents or trees or galaxies do; the only thing we know in the whole universe that has this feature is ourselves, and we're not even sure about each other—that's the problem of other minds.

Now, we are, in a sense, artefacts (and I mean that in the good sense of the term). We have been created by the process of evolution, both genetic and cultural. And what we're now trying to do is to reverse engineer ourselves, to understand what kind of a machine we are that this can be true of us.

Sue   **Are you equating subjective experience with having a point of view?**

Dan   Yes, but having a point of view is not a simple matter. There's an easy sense of having a point of view where lobsters have a point of view, and mosquitoes have a point of view. With a little stretching and pulling you might even say that a pine tree had a point of view; that is to say a pine tree responds to the world selectively—there's only some features of the environment around the pine tree that it's sensitive to and the rest of the world is indiscernible, as it were, by the pine tree.

But that's indiscernible 'as it were'. In our case there's 'real discerning'; and 'real discerning', in the eyes of many people who have thought about this, has got to be worlds away from the sort of discriminative capacities of that pine tree or that mosquito.

This creates an artefact in the bad sense of that term. To many people there's an imaginative chasm between us with our 'real discerning' and our 'real points of view', and the mere robots, or discriminating-but-not-sentient things. I think that the gap between me and a pine tree, or me and a mosquito, is huge but it's traversable by a series of steps. But I do have to say that some of the steps are quite counter-intuitive, and there's not yet in place the sort of firm 'take it or leave it' science that can force people to abandon their intuitions.

Right now it's a struggle to get people working in consciousness even to think about abandoning their intuitions. They have these powerful, seductive intuitions about how it has to be, or how it can't be, that are just wrong. Nothing new there! We've always had false intuitions about the way the world is, and counter-intuitive science has come along and changed them. But in this case, we don't yet know which intuitions to abandon and why. So the problem is very much a problem of persuasion and self-persuasion and a sort of self-manipulation of one's own imagination, which is scary to many people. So instead they try to have a theory which doesn't require them to tweak their intuitions at all, and then they end up down one cul-de-sac or another, because the theories that are not counter-intuitive are just wrong.

Sue I imagine that you may be thinking here of the zombic hunch?

Dan Yes. The zombic hunch is the idea that there could be a being that behaved exactly the way you or I behave, in every regard—it could cry at sad movies, be thrilled by joyous sunsets, enjoy ice cream and the whole thing, and yet not be conscious at all. It would just be a zombie.

Now I think that many people are sure that hunch is right, and they don't know why they're sure. If you show them that the arguments for taking zombies seriously are all flawed, this doesn't stop them from clinging to the hunch. They're afraid to let go of it, for fear they're leaving something deeply important out. And so we get a bifurcation of theorists into those who take the zombic hunch seriously, and those who, like myself, have sort of overcome it. I can feel it, but I just don't have it any more.

Sue What do you think the nature of this fear is? And, more personally, did you once have this fear yourself and have to overcome it? Or was it really quite easy for you to see that you shouldn't succumb to some slight desire to fall into the zombic hunch?

Dan Well let me start with that first. It wasn't a momentous occasion for me when, as an undergraduate, one day it just hit me that 'Oh yeah, Alan Turing had the basic move that we could replace Kant's question of how it was possible for there to be thought, with an engineering question—let's think how can we make a thought come into existence. Oh, we could build a robot. And what would it be for a robot to have a thought?'

So, resolutely, from the third person point of view, you sneak up on consciousness from the outside, not from the inside. You keep looking side-long at the inside, all along the way, and seeing if you can make the difference evaporate. There are powerful reasons for thinking that of course you can make the difference evaporate eventually, because it's got to; because we're part of the physical world; there's no mystery stuff; dualism is hopeless. So, since dualism is hopeless, let's see if we can figure out what the sufficient conditions are in purely material terms for there to be something that it is like something to be; something that has an inside; something that has a subjective point of view. And once I had that project clearly in my head, then it all fell into place. Now the question was just working out the details.

Sue **But you implied that sometimes the zombic hunch does tempt you...**

Dan Oh, it doesn't just tempt me. I deliberately go out of my way, every now and then, to give myself a good instance of the zombic hunch. I talk to myself, 'Come on Dan, think about it this way. Now can you feel it?' Oh, I can feel it all right. It reminds me of how you can look out on a clear night and, if you think about it right, and look at the sky and sort of tip your head just so, you can actually feel the earth in its orbit around the sun. You can see what your position is, how the earth is turning, how it's also in orbit, and it all sort of falls into place. You think 'Oh, isn't that quaint?'

This is a lovely perspective shift, but it takes knowledge and some very specific direction of attention to get into that frame of mind. Well, I think for people who have the zombic hunch and don't know how to abandon it, they have to learn to do something like that too. But they just haven't tried, and they don't want to.

Sue **Why don't they want to? What is this fear of letting go of the zombic hunch, even for people who might rationally understand the arguments for getting rid of it?**

Dan I think they're afraid that zombies would have no moral significance. Zombies would be just stuff, and you can chop stuff up, break stuff up, throw it away, burn it, whatever. It doesn't make any difference; it's just stuff. Whereas if we have immortal souls, or anything that's the moral equivalent, then we preserve our moral point of view. I think the idea of a soul is a curious fossil trace of the desire to treat ourselves as absolute.

Sue **Is it just about morality and mattering; that it matters to something or someone what we do? Or is it also about continuity—that we want to survive?**

Dan Well, I think those two are intertwined. Darwin made a great inversion of reasoning when he realized that you can have a bottom-up theory of creativity: that all the wonderful design that we see in the biosphere could be the products, direct and indirect, of a mindless, purposeless process. This simply inverts an idea that I think is as old as our species. It's what you might call a top-down theory of creativity: that it takes a big fancy thing to make a less fancy thing. Potters make pots; you never see a pot making a potter, or a horseshoe making a blacksmith. It's always big fancy, wise, wonderful things making lesser things. And so here we are; we're pretty wonderful, and so we must be made by something more wonderful still. I think it's very scary for people to give that up, and to begin to think about how our importance doesn't depend on the importance of something still more important.

You know, a good bumper sticker recipe for happiness is, find something more important than yourself to think about, but there are many such things that can replace the one big, important thing which many people think they have to have, which is God.

Sue I assume you don't believe in God. Do you think anything of the person survives physical death?

Dan Well, of course, many of the effects of a person's words and deeds can reverberate through human culture for some time after their death, and these can in rare cases be remarkably powerful and coherent. Abraham Lincoln is a more familiar presence *today*, better known, more recognizable, and more often thought about now, than most of the people who are actually alive today. I think that many people would *love* to have that sort of 'immortality' of effect, and would happily trade it for the more traditional prospect of a disembodied eternity in 'heaven'—an idea whose popularity is matched only by its incoherence.

But competition for admission into that pantheon guarantees that only a tiny minority will ever enter it, since the attention span of human culture is strictly limited. I wonder what the maximum value of p is, where p is the population of 'recognized immortals'. 1000? 10,000? When Elvis Presley finds his seat, does this force Dietrich Buxtehude out? That's the only sort of life after death, and it is in short supply.

Sue What do you think is your greatest contribution to consciousness studies? After all, the field has grown enormously since 1991 when you published *Consciousness Explained,* and consciousness studies has become all the rage. Where do you see your own contribution fitting in?

Dan I think, oddly enough, perhaps my most important or influential contribution was showing people that materialism was harder than they thought—that it was more counter-intuitive than they thought. Some of the reactions to that book have fascinated me, such as the people who've come up to me and said, 'I thought I was a good materialist until I read your book, and then I began to get really queasy because I realized I have to give up a lot more of my intuitions about consciousness than I thought.'

I said 'Absolutely right! You have to embrace the counter-intuitiveness of some of these ideas. You can't just trust your common sense. There are some deeply disturbing aspects of any proper materialist theory of consciousness. So let's get on with it and expose them.'

One of my favourite sequels to the book, is that a lot of subsequent work confirms that I'm right. I think I was pretty much the first to put forward things that have now become well established phenomena, like change blindness, which I predicted. At the time, this provoked outrage or frank disbelief. People said, 'You're out of your skull there,' but I said, 'You wait, you'll see,' and sure enough, the effects are real. In fact they're much more potent than I dared claim. I sort of wish I could go back and put a little more vim into some of my statements there because, in retrospect, I was more cautious than I should have been.

Sue Can we take change blindness as an example there? I think that if you take the findings seriously you have to wonder about every act of vision you make. All the time, in your everyday life, looking around you, you have to realize that you're conjuring something out of nothing; that you have far less information in your head than it seems. It should, and for me to some extent does, change the way you feel about your role in the world.

Does it have that effect on you personally? Has predicting change blindness, and then realizing it was an even more powerful phenomenon than you'd thought—has that changed for you what it's like being Dan Dennett alive and looking around the world?

Dan I wish I could say 'yes', but in fact I think the answer is 'no'. I was thinking about those things even when I was an undergraduate. Here's another way of looking at what my contribution has been: consciousness looks like an insoluble mystery when you have an inflated vision of what consciousness is, and our introspective lives tend to give us that inflated vision. We tend to think we're conscious of a lot more than we are; we tend to think that consciousness has properties that it just doesn't have. If it did have those properties, boy oh boy it would be much harder to explain than it is. So the first thing you have to do is deflate the phenomena so that you can see that they're not quite so gosh-darn wonderful—so truly mysterious—as you thought they were. Then they're sort of tamed. Then we can explain them.

Of course, there's tremendous resistance to the deflationary move. People don't like me saying that they're not conscious of as much as they thought they are, and what they are conscious of doesn't have the features that they say it has. Their reaction to this is 'Oh Dan's just denying the existence of consciousness.' No, I'm not. I'm just saying it's not what they thought it was. Now it's interesting if you look at the history of science. The term used for talking about the pre-theoretical catalogue of the properties that have to be explained was the 'phenomenology'. So Gilbert worked out the 'phenomenology of magnets', for instance. These are the phenomena, this is what has to be explained. As for the 'phenomenology of consciousness'; if you are an auto-phenomenologist, if you are an introspectionist, if you adopt the first person point of view, you're just going to get it wrong. You're going to con yourself into supposing that your consciousness has many features it just doesn't have. So the trick is to characterize the method which neutrally categorizes the phenomenology of consciousness, and then, go to work. Explain it! And when you've explained all the phenomenology, you're done. You've explained consciousness.

Sue   And is that what you call heterophenomenology?

Dan   That's heterophenomenology. Heterophenomenology is the scientific catalogue of what has to be explained.

Sue   You're very hard there on the first person point of view, but can you see no role for disciplined self-observation? I'm thinking in particular of meditation, where it's said that if you keep practising long enough, some of these things become obvious. The visual world starts to disintegrate. Our illusions of the continuity of self, the continuity of the perceived world, the simultaneity of events, all start to fall apart and you see more clearly. Do you think there's any truth in that, or do you dismiss that completely?

Dan   No, I think there is truth in it, but this is in the context of discovery not the context of justification. Every experimenter should, of course, put herself in the apparatus, and see what it's like from the inside. You should certainly treat yourself informally as a subject and see if you've overlooked something, for instance. But having done that, then you do the experiment. You use naive subjects, and you figure out some way to get what you've discovered from the first person point of view to manifest itself for neutral observers from the third person point of view. And if you can't do that, then you have to be suspicious of the insights that you thought you had.

In a sense this is obvious. Nobody in the scientific world working on consciousness would think of submitting a paper that said, 'Well I introspected under the following circumstances and these are the things that I thought.' If you think you've discovered a phenomenon, then you go out and test it using the scientific method, and that means the third person point of view.

This could be just a typical philosopher's hypersensitivity to form and rigour, if it weren't for the fact that so many people are just wrong about the results of their own introspection. People cannot prevent themselves from theorizing when they think they're observing.

Sue   One of the things that's amazed me over the years, is how systematically and deeply you are misunderstood. I'm thinking of things like heterophenomenology, the third person perspective, the zombic hunch, the Cartesian theatre. Do you understand why people find them so difficult? You write clearly; you explain things well, at least I think you do. How come you get so misunderstood all the time?

Dan   Well, I wish I knew. I wish I knew. I've got hunches about it, and here's what I think happens. I've caught myself doing this with others, so I can see how they can do it with me too. When somebody tries to tell you something which is initially very counter-intuitive for you, you put your best effort into it and then translate it into your own terms, so that you can understand it. So you're not just listening cold, you're actively translating what you're hearing into your own dialect. But of course this can horribly backfire. If somebody is trying to put forward something that really is counter-intuitive, you almost certainly get it wrong. You'll throw out the most important part and you'll turn it into one kind of nonsense or another. And if you're not alert to that, you'll think 'Look, I did my level best to understand this person, and here's what I come up with. That's just crazy, so she's just crazy.' Nobody wants to hear that maybe your level best wasn't good enough.

I also think that in a way my writing style traps me, because at least superficially its not hard to see what I'm doing. It goes down quite smoothly, not like reading Hegel or Heidegger. So people think it's easier than it is. No, it's actually very hard. I tried to make it as easy as I can, but it's still very hard, and if you make the mistake of thinking that it's actually a pretty simple idea or two, you're just wrong. But I can see why people would think that.

Sue I'm particularly interested in one of your central arguments: the non-existence of the Cartesian theatre. You explain why there can't be a Cartesian theatre in the mind or the brain, why there isn't a show going on in the head, and why there isn't somebody watching. You call people who think they are materialists but are still trapped in imagining a Cartesian theatre, Cartesian materialists. Can you say something about what you take to be the signs of being a Cartesian materialist, and how common they are?

Dan The sure sign of Cartesian materialism is anybody who tells the story of consciousness and doesn't go on to answer the 'and then what happens?' question. So it's as if we work so hard to get this stuff up and presented to the Queen and then what happens? We get inside the audience chamber but why does that make it consciousness? Any theory that's still got a place for the show to happen has not yet done the job.

A curious feature of this is that if you then go on and answer the 'and then what happens?' question, a lot of theorists are sure that you've left something out. Because now we're back explaining behaviours and reactions, and the effects on vocalization and memory, they want to say, 'Wait a minute, where's the consciousness?'

There's a bi-modal distribution between people who think that any theory of consciousness that leaves out the first person is a hopeless theory, and those who think that any theory of consciousness that *doesn't* leave out the first person is a hopeless theory. You've got to leave the first person out of your final theory. You won't have a theory of consciousness if you still have the first person in there, because that was what it was your job to explain. All the paraphernalia that doesn't make any sense unless you've still got a first person in there, has to be turned into something else. You've got to figure out some way to break it up and distribute its powers and opportunities into the system in some other way. So the Cartesian materialist is the one who describes large parts of the machine, but it's still inhabited.

Sue I think I see it in things like, 'and then it's displayed', and 'then it enters consciousness'. Would you count those as signs of being a Cartesian materialist?

Dan Those are certainly danger signs, unless the person goes on and cashes that out very carefully.

Or there might be a theory that says, 'And then our brains tell us *blah blah blah*'. So who's this 'us'?

Sue I like your idea that you deliberately throw yourself into the zombic hunch. Perhaps I should sometimes throw myself into the Cartesian theatre more willingly than I do.

I do sometimes get into it. I can get quite upset thinking about—say— the brownness of this desk here; the 'how it is to me'. I have the very powerful conviction that I am in here experiencing this ineffable, unique, private sensation of the brownness. Can you help me? I know quite a bit about your theory, but when I get really badly into that feeling, how can I get out of it?

Dan The way I recommend is to ask yourself, 'What am I pointing to? What am I ostending when I say *this*?' What I think you'll find is that you can start elaborating a sort of catalogue of the facts that matter to you at this moment. Maybe it's the particular deliciousness of this taste in my mouth; so what is that deliciousness? Well I'd like some more, and I can recall it at a later date, and so on. We're going to take care of all that. We're going to include your disposition to want some more, your capacity to recollect, and even the likelihood that you will find yourself pleasurably recollecting this experience of it. There's a huge manifold of reactive dispositions that you're pointing to when you're saying 'This very yumminess right now,' and what you have to do is recognize that however indissoluble, however unanalysable, however intrinsically present that all seems to you, what has to be explained is that it *seems* to you, not that it is so.

Sue But you have said there that it *seems* so to me. Presumably you would say that we not only have to explain why it seems so to me but why it seems there is a me to whom it seems.

Dan Yes, exactly. Those are the two halves you've got to explain.

And people—wonderfully conveniently for them, and inconveniently for the truth—forget that it seems that way to the zombie too.

Sue Do you think you have free will?

Dan Yes.

Sue And what do you mean?

Dan I mean that in all the things that matter to me I can make a decision based on my consideration of what matters the most and why. I wouldn't have free will if I were obsessive, or were an addict, or were seriously deranged so that I couldn't keep track of reasons, or if I had a memory disorder that meant that I couldn't keep track of a project from minute to minute. Then my free will would be pointless.

The model that we want to have for free will is of an agent that is autonomous, not in some metaphysical sense, but in the sense of being able to act on the reasons that matter to the agent, and who's got the information that is needed to act in a timely fashion. In order to appreciate this you have to realize what brains are for. Brains are for generating expectations about the future. The simplest imaginable thing a brain is for, is for ducking an incoming brick. You see the brick coming. You see it's heading for you. You expend a little energy to duck so it doesn't hit you. There's a lot of things to avoid in life; and there's a lot of things to try to accomplish, but let's take the verb 'avoid'. It's key in one particular regard, that the word inevitable comes from it. Inevitable means unavoidable, and it doesn't really make sense in a context where there isn't avoiding. Where there's avoiding, there are things that are inevitable and there are things that are, if you like, 'evitable'. What makes it possible to avoid things is having some foreknowledge of what's going to happen. So, that's what our brains are for.

If you've got that equipment and it gets used, you can have reasons for acting that are good reasons, that are your reasons. You didn't make them out of nothing, you made them out of all the information and all the values that you've ever considered and reflected upon and decided upon, and for better or for worse, you've come up with a particular set of values, and now you're ready to act.

Just take a simple case of a chess player who makes a move. Why did he make that move now? I might say, 'Well the clock was running, I had to decide sooner or later. OK, enough thinking, it's time to move. This is my move. It may not be the best move. I may live to regret it. I may discover a better move in a few seconds. I may get checkmated. But I wasn't deceived about the position of the board. I wasn't deceived about the rules. I wasn't deceived about the point of the game. That was the best I could come up with. So that's my free will, that's my move.'

Sue In all of that description of free will, you kept saying 'I, I, my, my'. I want to know how you relate this to your ideas that there's no audience in the Cartesian theatre, that the self is a benign user illusion and so on. Who is it who's having the free will?

Dan   The agent.

Sue   By 'the agent' do you mean the whole body?

Dan   Sure.

Sue   Then isn't it important that you distinguish that view from what many people feel about free will, which is that *they*, the audience in the Cartesian theatre, the little special conscious me inside here, is the one that has the free will?

Dan   One of the most curious ironies to me, in my earlier writing on this, is that the most important sentence in my book *Elbow Room*, I put in parentheses, and so nobody paid much attention to it. I said—and it was meant ironically—that if you make yourself really small you can externalize virtually everything.

The imaginative pressure to think of yourself as very small is easy enough to find. When I raise my arm, well what is it? There must be some part of my brain that is sort of sending out the signal and then my arm is obeying *me*, and then when I think about the reasons why, it's very natural to suppose that my *reason store* is over there somewhere, and I asked my reason store to *send* me some good reasons. So the imagery keeps shrinking back to a singularity; a point, a sort of Cartesian point at the intersection of two lines and that's where I am. That's the deadly error, to retreat into the punctate self. You've got to make yourself big; really big.

One aspect of this has been very nicely expressed in recent years by Andy Clark, in *Being There*, that we offload a lot of our minds into the world, we then do our thinking using those peripheral devices as part of our equipment. We don't have to do it all in our heads, we can do it out there with slide rules, or calculators, or laptops, or with a little help from our friends.

In fact I think most of us who manage to live moral lives, lives that we're not ashamed of, in fact rely a great deal more on the support of our friends than we readily acknowledge.

Sue   You told me that you had many of your ideas as an undergraduate, and that what you've been doing since has really been fleshing those out and explaining them, but I'd like to know this: has anything happened in your life as a philosopher of consciousness that's really changed you, or changed the way you feel about yourself?

Dan I haven't had any conversion experiences that I can think of! But I certainly think my interactions with people outside philosophy have had a huge effect. Probably the first five years of my career, back in the upper Neolithic and the late '60s, I was still hanging out most of the time with philosophers, and a rather small, but cherished and fascinated percentage of the time, I spent talking to people of other disciplines. Gradually I came to realize that I learned more, and found it philosophically more interesting, talking to people about artificial intelligence, biology, neuroscience, and psychology, than I did talking with my fellow philosophers. So over the years I went where the fun was for me, and it fed on itself. I got invited to more and more non-philosophical occasions and conferences, and I read more and more articles and books, to the point now where I read the philosophy as a duty. It staggers me to realize how much less fun it is to read most philosophy than it is to read good biology or good psychology or good artificial intelligence. And so that's made a huge difference to me.

Of course, this means that a lot of people in the field of philosophy say, 'Well you know Dennett's just no longer a philosopher, he may have been a philosopher once but he isn't now.' Well, I don't want to argue about that, but if so, then maybe all philosophers should cease being philosophers and try to do the sort of thing that I'm doing, because I think I'm getting philosophical results, and making philosophical progress. I think it's far better than the sort of vacuum philosophy that we all used to engage in back in the old days.

Sue And what *is* philosophy?

Dan Philosophy is what you do when you don't yet know what the right questions are to ask.

Susan Greenfield

*I get impatient when
the really big questions
are sliding past*

Baroness Greenfield (b. 1950) took classics at school but changed to psychology and physiology at Oxford. She then did a DPhil in pharmacology at Oxford before becoming lecturer and then Professor of Pharmacology there. In 1998 she became Director of the Royal Institution of Great Britain, and in 2001 became a Life Peer. Her research concerns neuronal mechanisms and degeneration in Alzheimer's and Parkinson's disease as well as the brain basis of consciousness, and she has founded two neurotechnology companies. Her books include *Journey to the Centers of the Mind* (1995), *The Private Life of the Brain* (2000), and *Tomorrow's People* (2003).

Sue I know you've described consciousness as 'one of the last great mysteries of science.' What's so mysterious about it?

Susan The fact that it's a subjective phenomenon that we can't really define properly. Everyone knows what it is, but we can't use the normal operational definitions for defining it; and therefore it's very hard to know how to even frame the question as to how a subjective inner state is associated with something physical.

Sue Wouldn't that lead you quite easily to become a behaviourist and say therefore we shouldn't even try?—I mean, if you say we can't define it, can't pin it down...

Susan No, on the contrary I think what you can do, and what I've attempted to do, is to establish correlates—and I use that word cautiously; that is to say, even though you can't establish a causal relationship between one thing and another, a start is to see how the two co-vary. And I think drugs, for example, are a very good way of looking at how the two things co-vary, because drugs can modify consciousness, they can even take consciousness away, and at the same time we can actually document and quantify and explore how drugs work on the physical brain.

Sue So could you give me just one example of a specific drug and a specific effect on consciousness?

Susan Anaesthesia, for example, which takes consciousness away. This is something I'm working on at the moment, actually. We know that it can't be localized in a brain region, and certainly not in a gene and certainly not in a chemical; so I think it's a very nice way of studying consciousness because it forces you to be, if you like, in meta-space. But by the same token, my great insight for myself was in exploring more about anaesthesia and finding out you get levels of anaesthesia—that led me to think, well, if you have unconsciousness in terms of degrees then you could have consciousness in terms of degrees as well. Therefore one can approach consciousness as something that can be quantified, that's not qualitative, that is more tractable to science—because as you know science is in the business of quantification not qualification. So what one can do is to look at degrees of consciousness and then you can go into the brain and actually look and see what is in the brain that varies from one moment to the next.

Sue But if you're talking about measurement, typically with anaesthetics one might take the standard scales that measure depth of anaesthesia— but these are not really getting at consciousness as you've described it, as subjectivity. So how can you get at real correlates between consciousness and anaesthesia?

Susan Well, at the moment the imaging is sub-optimal, but what you need to look at is to find something that is actually a true index of consciousness rather than being something that is necessary but not sufficient for it.

Sue Ah, but could there be such a thing as a true index of consciousness; because if, as you said at the beginning, it's a mystery because it's subjective—because it's what it's like from the inside—how can you have a true index of that?

Susan Well, I think you can have an index of something without necessarily *it* being the thing itself. So what I've suggested is that, for example, one can have an assembly of brain cells, and the size of that assembly will correlate with the size of consciousness, yes? That is not to say that if you take an assembly of brain cells and put it in a teapot it's going to be conscious—of course not. It's an index; it's a bit like the monitor light on your iron—when it's on, that's an index of the iron being on, but it's not an iron in itself. It's an index of it, you see?

At the moment with anaesthesia, although just looking at a whole constellation of pulse rate, heart rate, pupil dilation, and so on, will tell one that probably someone's anaesthetized, that is not in and of itself the final parameter one looks at; and I think that looking at assembly formation that might be the parameter. But at the moment clinical imaging is vastly too slow to actually capture what's going on.

I like to think that modern brain imaging is a bit like those old Victorian photographs, where you can see very valuable things that are exceeding a certain time frame of exposure, but not the interesting things that occur swiftly within it. So yes, you can see a brain tumour, you can see a steady state, but what you can't see is the transitory formation of an assembly; in fact we know that tens, hundreds, and millions of brain cells will corral up in a quarter of a second and then go again.

Sue You have argued that consciousness gets bigger with the size of the assembly—or deeper, or wider, or whatever is the appropriate word. What led you to that idea, and what would count as evidence against, or for, it?

Susan OK, well let's look at the other candidates. I think you and I would probably both go along with the assumption—and it is an assumption: we're not pan-psychics, or I'm not and I don't think you are. So let's assume that consciousness is generated by the brain; in which case let's look at the candidates in the physical brain.

Now, could it be the genes? Obviously not; genes don't have the gene of consciousness, and genes just make proteins, so certainly I wouldn't think of that very seriously. Are there 'chemicals of consciousness'?—that's what some people like to posit; what they mean by that is a shorthand for 'there are chemicals that modify consciousness'—that's not to say that chemicals have consciousness inside them. At the other end, are there brain regions for consciousness?

No, there's no such thing as a centre for consciousness; and we could rehearse all the arguments, but I'm sure Dennett has already, and we all know them, I'll take that as read. So we're starting to run out of options here, yeah? So if you then look at the hierarchy of brain organization, the only thing left is the middle level between the chemicals and synapses and proteins and between the macro brain regions; and that is the level which is actually the most dynamic: that of neuronal networks.

Sue So would you say that neuronal networks generate consciousness?

Susan No, I wouldn't. I would say that they are a sensitive index of it. As I said, if you took an assembly of brain cells and put them in a teapot they wouldn't make consciousness, which is why I get slightly irritated when people who work on brain slices look with great glee at their 40 Hz oscillations; of course 40 Hz oscillations might well be a necessary quality of assemblies—but as I said to John Searle, there's a difference between necessity and sufficiency. He said, well, there's something else as well—and I said, 'Of course, it's the something else that's important, yeah?'

Sue What I'm really trying to get at is the question of what you mean by *generates* consciousness.

Susan It's a correlate of consciousness.

Sue But there's a big difference, and you did use the word 'generated'.

Susan No, only correlate, because, as I think I said at the very beginning, if you'd said to me you'd found out how the brain generates consciousness, I don't know what answer I would expect. Would it be a formula, would it be an experiment, would it be a subjective experience, would it be a model—what kind of thing would it be that would satisfy someone that you had discovered how the brain generated consciousness? If you said to me you'd built something that defied gravity, I'd know what kind of thing I'd be expecting.

I do find that hard—and also what would we do or know that we don't do or know now, if we knew that? So even the very question 'how does the brain generate consciousness?' is not a specific enough question, and it's one that at the moment has been mutilated by different scientists—Koch and Crick and so on.

Sue They certainly think that the brain generates consciousness, but some philosophers would argue that that's entirely the wrong way of looking at it; many functionalists would say that the brain doesn't generate consciousness at all, that it generates intelligence, vision, and all these processes, and that that's all there is to it; that there's not something else called consciousness. So I'd like to know whether you are a functionalist in that sense, or whether you think that consciousness is something separate from all those processes.

Susan No. This is actually one of the issues that came up when I used to teach at Oxford; I was teaching vision one day, and we'd plodded through all the brain areas, and the Hubel and Wiesel work, and then I said, 'So how do we see, then?' And they said, 'Oh, that's consciousness, isn't it; is that on the syllabus?'

No, my own view is that you can't say that you study the brain but you're not interested in consciousness; it's like saying you study the stomach and you're not interested in digestion.

Sue Can you separate consciousness at all from anything? I suppose the best way of asking that question is, Do you believe in the possibility of the philosopher's zombie? Could you have someone who looks exactly like Susan Greenfield, and speaks exactly like her, and has this same discussion with me, but it's all dark inside, there's no subjective experience?

Susan No, I think that consciousness is part of feeling, part of seeing; so I don't think you can separate out vision and emotion from consciousness, no.

But I always have problems with these philosophers' thought experiments; taken to the extreme they lose their value, but let's take them to the less extreme. For example, Sony have made this wonderful little animal, this man called QRio, which is better than Aibo the mechanical dog, and can actually have a conversation, a rather interesting, surreal, Pinteresque conversation; he says dreams are the most important thing … it's quite sweet actually. My own view is that of course you can build clever things that do clever things; but you can be conscious when you're not doing anything, when you're just lying there in a flotation tank. Things that move and talk like QRio can give a semblance of consciousness, and other people can seem to be utterly brain-dead but of course are conscious—like many of the people I know who just sit around. So therefore I think you can disassociate behaviours from consciousness.

Sue How would you know if you just had consciousness separate from behaviours?

Susan Well, you don't, of course; I mean, someone can be lying there with their eyes closed, and you don't know if they're asleep or awake.

Sue   But ultimately, if you had the sort of index you're after, you would then be able to tell?

Susan   Yes, yes, yes.

Sue   What about other animals? You talked about the size of neural assemblies; and I wonder how this affects which creatures could be conscious?

Susan   Well, again I disagree with many of my colleagues: for example, Gerald Edelman draws the line at lobster level, disenfranchises the lobster from consciousness, perhaps because he boils them and eats them or something. But my own view is that anything with any brain, however rudimentary, will have a degree of consciousness proportional to that. This also means that a foetus will be conscious, as soon as you get the brain in some way growing, yes? So it's like a dimmer switch: consciousness grows as brains grow, I suggest.

Sue   As they grow: do you mean just through the lifetime of an animal?

Susan   Ontogenetically and phylogenetically.

Sue   Both. So a bigger brain means more consciousness?

Susan   Not literally a bigger brain—more complex brain means more consciousness; because, as you know, certain brains like our own are not necessarily the biggest brains in the world, but we have the most convoluted cortex and so on.

Sue   So what is it then that matters: brain size, convolutions, the size of neural assemblies ... what is it?

Susan   It's a whole combination of physiological and anatomical features of our brains; size, convolutions of the cortex and therefore surface area of the cortex; and also the relative job that different brain regions do. It's a quantifiable thing.

Sue   So ultimately, when we know a bit more, you would expect to be able to look at any brain, whether it was a lobster, or a cat, or a fish, or a bird, or us, and say how conscious it was?

Susan   Perhaps yes, at the ultimate we'd be able to do that, and say the degree of consciousness.

Sue   You've talked about animals there, and you've mentioned QRio; what would you require to build an artificial consciousness on this view?

Susan   Ah, that's a slippery question: slipped in there is the fact that one can build an artificial consciousness.

Sue Please say you can't, if that's what you think.

Susan I don't. Well, put it this way: I get angry with people like Ray Kurzweil or indeed Dennett, who violently say you can; and some people lampoon me, or caricature my view, of saying you can't. Now that is a very non-scientific approach either way, because it's relying on faith rather than on reality.

A more, to my mind, open-minded attitude is to say to those people, yeah, not only is it a problem to build such a thing, but given that I don't know whether you're conscious or not how are they going to prove it anyway: how are they going to prove that that agent is conscious—given that I've dismissed the QRio model, yes? So my own view is not very helpful, because if you knew what it was that you wanted to build or that you wanted to prove, then you'd have solved the problem anyway.

So there are two problems: a) you don't know what you want to model anyway; and b) even if you did know what you wanted to model, that would itself solve the problem. So therefore to me it's a no-brainer—I can't see why people worry about it.

Sue Doesn't it have a moral issue about it, though; that if you built things that were capable of suffering, which many people feel is an intrinsic part of consciousness, you would have a responsibility that you don't have if you know that they couldn't be conscious? Doesn't that make it an important issue at all?

Susan Well, it's begging the question to say, yes, will it suffer or not? But my own view is that it's so unlikely—I'm sorry to sound so pragmatic, but it's like arguing angels on the head of a pin. I think frankly that if one said, 'OK, we've got a choice here, of building a robot like QRio'— and no one thinks QRio is conscious—'that can go into bombed-out buildings and rescue people who are dying and ill—but we're not going to do this, in case this machine might, on a million to one chance, suffer itself'; I know I wouldn't have a choice. I wouldn't have any problem deciding what I wanted to do.

Sue You wouldn't agonize about it for a minute?

Susan I would not, not for a nanosecond. It might be an interesting kind of philosophical question, but not a pragmatic one. The whole approach of artificial intelligence is very useful if indeed you are building things that will go into dangerous, unpleasant, or boring places where human beings can't go, but I don't think it's a very obvious route to understanding how the human brain, or any brain, generates consciousness.

Sue   Do you have free will?

Susan   That is one of the most interesting questions and one that I keep coming back to each time. I'm not such a fan of Searle's, but I'm quoting him a lot: he said that when he goes into a restaurant and orders a hamburger, he doesn't say, 'Well, I'm a determinist, I wonder what my genes are going to order.'

Sue   I do. You're right that Searle doesn't do that, but when I go in a restaurant, I think, 'Ooh, how interesting, here's the menu, I wonder what she'll choose'; so it is possible to do that. But what do you do?

Susan   I would say that, yes, I am under the illusion—possibly it's an illusion, but as we all know if you believe in the illusion it's not an illusion. Now, I think you have to make that choice, because a lot of other things follow: if you don't do that, what do you do with the criminal justice system? For example, if no one has free will, it means that no one should be in prison.

Sue   No it doesn't, because although you get rid of retribution you would still put people in prison in order to provide a deterrent for other people and in order to keep really dangerous ones off the streets—so some of the system would survive.

Susan   But how can it provide a deterrent for people if they don't have free will; it's not up to them.

Sue   Because part of the deterministic system is the deterministic effect of punishment and threats of punishment.

Susan   Well, I don't know if that is a deterrent. If for example you have—oh, I love this, 'the criminal gene' or an overworking area of the brain which is disposing it, or—as in the Twinkie defence, where someone makes somebody cupcakes, they became hyperglycaemic and committed murder on the basis of it. If for any of these reasons you know you're not going to be deterred, the fact that someone else was sent down for murder is not going to, by definition, stop you. So if you say we don't have free will, where should we draw the line between one person and another: was it Osama bin Laden's genes, were you predisposed, was Hitler? How would we feel if we were saying, 'OK, if you don't have free will it wasn't Osama's fault'?

Sue But surely you, as a scientist, shouldn't be saying, 'Ooh, there'd be terrible consequences if we believed this, so we mustn't believe it.' Don't you think truth is a higher goal?

Susan No, no, you've misquoted me on the second thing: I'm saying there may be terrible consequences; I haven't said 'Oh God, we mustn't believe that.'

Sue But you were implying that one's attitude to free will affects the way that you personally live your life, and that's what I'm trying to get at.

Susan Yes, and it makes me think a lot about it, because I'm very interested in the way society is going, the way people apportion blame to people for things; and as we deconstruct in our ham-fisted way the human brain into plasticity and genes that make the proteins and how they're switched on and off; and as we get better at brain imaging, and see bits of the brain lighting up—well, my own view is that those things are massively over-hyped in what they mean, massively, yeah? Nonetheless, it does give some people the illusion that if you can deconstruct the brain in that way then you can come up with the reason for something, and that therefore the reason for something is not the person's fault. Now that is what concerns me hugely in this society, as scientists make more and more hectic claims on where we're heading: what will be the implications on how we view the individual and the sense of responsibility they might have for themselves? And it's all of life, not just criminals—how much a kid at school feels they're responsible, and feels their destiny is in their own hands, and how much it's because of all these influences. But the only thing that concerns me is that everyone's treated the same, that we don't have groups of people who are ring-fenced and pampered because it's not their fault, and with others it is their fault and therefore the full weight of justice comes against them.

Sue But you seem to be coming close to saying, 'Well, I know it's an illusion really, but we ought to have everybody believing in it because otherwise ...'

Susan No, I feel it's true. Without sounding too exaggerated, the whole of reality is possibly an illusion. So yes, in one sense everything's an illusion; but on the other hand, I believe very much in my own free will. So I can see that you might be, in your Sue Blackmore way, sitting there and saying, 'I wonder what she's going to order' and so on, and that might be quite fun; but I don't think that every minute of your life you think, 'I wonder what she's going to do.' Well, you might if you have schizophrenia, but I think for most people most of the time, you have to assume that other individuals are acting of their own free will, and that you yourself are a cohesive entity.

We all know the Libet experiments; I myself have been the subject of one of those, where prior to wanting to do something your EEG's changed already. That doesn't threaten me at all, and it doesn't mean to say that some ineffable me is being controlled by my brain, as some people bizarrely think; it just shows the workings of my body going on.

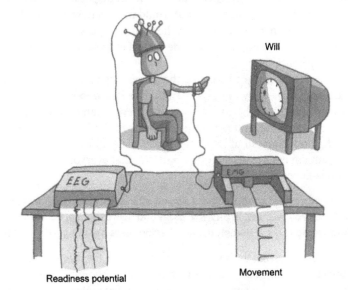

Readiness potential

Movement

Will

In Ben Libet's 1985 experiments, subjects had to flex their wrist spontaneously and deliberately at a time of their own choosing. The time of the movement was measured using EMG (electromyogram) electrodes on their wrist; the start of the readiness potential in their motor cortex was measured using EEG (electroencephalo-gram); the moment at which they consciously decided to move was measured using a spot revolving on a screen; they had to say (after the movement) where the spot was at the moment of willing. The results showed that brain activity began nearly half a second before the will to move. Libet's controversial experiments have been interpreted by some as having implications for free will.

Sue **When you talk about 'my body', you might imply there's a 'you' and there's a body.**

Susan Well, OK, I can say 'the body of me', or 'the body that's called Susan Greenfield', or whatever semantics you want to use.

Sue So you don't think you are separate from your brain; presumably then you don't believe in any kind of life after death?

Susan No, but I don't have some kind of conviction and zeal against anyone who believes in anything that has an element of belief, or faith, or religion in it. My own view is that a true scientist is open to all ideas until they are disproved.

Sue So you'll wait until your own death to decide about that one, will you?

Susan No, not necessarily. Given the facts at our disposal at the moment, I can't see how it could happen; it would mean a new type of physics. Given that for me the personality, the brain, the person, the mind, and so on is so intimately affected in the brain, I can't see how all this could exist without a brain.

Now just because I can't see that, does not mean to say it is not true, and it certainly doesn't mean that people who have very strong beliefs that this is the case are any less clever than me; so I'm not as arrogant as some, and at the end of the day I think I'd like to keep an open mind: I myself at the moment can't see how it would be possible, but I'm not going to say emphatically that everyone else is wrong who believes in that.

Sue You've studied consciousness on and off in connection with all the varied work you've done, in pharmacology and neuroscience, for a long time; how would you say that studying consciousness has changed your life?

Susan It's interesting, because in a way I suppose I've always studied it, since I did classics at school. In a funny way, although I didn't realize it, I suppose I've always been interested in what makes a person a person, the issues of free will which change from Aeschylus through Sophocles to Euripides, from being a kind of determinism to being a kind of individual internalization of decisions. So in that sense I was introduced to these ideas very early, by my very enlightened school teacher. I think I didn't come to this as a scientist; it was all the other way round: I was someone who was already fascinated by philosophy and those big questions, who then saw in science a medium for approaching it.

Sue But having learnt all that you've learnt, and developed different theories about it; has that changed the way you live your life?

Susan It has. It's certainly changed my attitude to science, in that I get more impatient with what I call science accountancy, and the i-dotting and t-crossing, and the almost anal attitude to some ways of doing science, when life is so short; it's like rearranging the chairs on the deck of the Titanic when the big questions are sliding past—sorry to mix metaphors—while people are fretting about a receptor sub-type.

So in that sense it's something that sets me apart from other scientists, in that I tend to get impatient with that type of stuff, and more depressed than many are about how much we know about the brain. I feel that we're just really exchanging anecdotes with each other; no one has done what they've done for physical sciences: come up with a proper framework that everyone buys into, with laws and rules and principles and so on, that successfully brings together the different levels of working on the brain. So in that sense I feel a frustration when people are so complacent and pleased with themselves, and go to big meetings on the brain, where everyone pats each other on the back because they've done so well, when really, I think, we're at the very very beginning.

Sue And as this great ship slides on by, what's the really big question for you?

Susan Well, how the brain generates consciousness.

Richard Gregory

Science is full of gaps

Richard (b. 1923) served in the RAF during the Second World War and then went to Cambridge to read philosophy and experimental psychology, where he remained for many years, directing the Special Senses Laboratory, investigating the recovery of a completely blind man, and beginning his work on visual illusions and the idea of perceptions as hypotheses. In 1967 he founded the Department of Machine Intelligence and Perception in the University of Edinburgh and worked on early robots. Then from 1970 he moved to Bristol where he was Professor of Neuropsychology and Director of the Brain and Perception Laboratory, and where his love of science and asking questions about everything led him to found the hands-on science centre, the Exploratory. Among his many books are *Eye and Brain* (1966), *Mind in Science* (1981), and *Odd Perceptions* (1986). He is editor of the *Oxford Companion to the Mind* (2004).

Sue Tell me **what you think the problem of consciousness is ... why is consciousness such a problem?**

Richard The real problem, though it's trite really, is the huge gap between what qualia are like and what the physical system of the brain is like. In other words, how the hell does physics produce something which is so totally unphysical?

But then I turn round on myself and I say it's not really a problem at all because science is full of gaps. Let's take electricity produced by a magnet moving through a coil of wire, which Faraday found in 1831. You wiggle this magnet and, blow me, there's something utterly different happening: electricity! So perhaps the problem is just emergence like electricity.

Sue So you seem to be veering between two views: sometimes you think there really is a horrible gap and you don't know how to think about it, and at other times you say 'Hey, it's only a gap like every other gap.' Can you explore that feeling a bit more with me?

Richard Well, I used to think that the appearance of emergence is a sign of our ignorance. If you had an adequate model then you could fill in the gap, you could walk up a conceptual ladder from the model and see how the phenomenon arises. Then the emergence would disappear. I think I still think that, though I'm not quite certain.

Sue Does that lead you to a view similar to Francis Crick's, that we'll ultimately succeed and therefore the best thing we can do is to get on with the brain science and wait until the gap is closed?

Richard Yes, except that it may not come from thinking about brain science in the way we're thinking today.

Sue Oh, what else might it be?

Richard Well it might come from anything. I mean in the history of science these gaps are often filled by some incredibly indirect means, and you don't necessarily get there by the obvious route. Take the discovery of X-rays around 1900. At first this looked totally mysterious, and then it became explicable as just one wavelength in the spectrum, so it wasn't a huge gap at all. It might be analogous to that.

Sue That's rather an exciting thought; that we might make a totally unexpected discovery, and meantime here we all are, squelching around in the mud of ignorance, not even knowing what to look for.

Richard Yes, that's kind of how I see it. I don't think there are guiding principles. The big gap is really a sign that you don't know where to look for the answer.

Sue Can you remember when you first became interested in the problem of consciousness?

Richard I've been interested in perception for about a million years but I didn't actually think much about consciousness simply because I didn't know how to think about it. Then I did an article in the *Encyclopaedia of Ignorance* about 25 years ago.

It was quite funny because the publishers wrote to me and said they'd got cold feet about the title, and I said, 'Well I don't want my article on consciousness published under any other title because I know I'm ignorant and it's ideal publishing under the rubric of ignorance.' I suppose others must have said the same thing, because it didn't get changed.

Sue But didn't you think about it before? You read philosophy at Cambridge with Bertrand Russell and other famous people. Didn't you think about the mind-body problem then?

Richard We certainly thought a lot about the 'other minds' problem, which John Wisdom went on and on about. He would ask, for example, 'Is another mind like a fire on the horizon?'; this might go on for weeks. But what we didn't talk much about, as I remember, was the relation between brain activity and consciousness; really because we didn't think much about brain activity at all. This was so even for psychology at that time—which as a matter of fact is the reason why I moved into artificial intelligence. It seemed too difficult to do the physiology on the brain.

And as for consciousness, I think the trouble was that we hadn't a clue what to say that might be worth saying. It's a little bit like the frogs that die of starvation unless things are moving around them. If they've got all the food in the world, and it's not moving, they can't see it and they die of starvation.

Or it's like playing a game isn't it? I mean, I'm quite good at table tennis so I can enjoy thinking about how I could improve my game—or chess. But if you've got a problem like consciousness, and you haven't got a clue how to tackle it, you don't think much about it because it's a waste of time. I don't like contemplation much. I like taking a problem and trying to solve it.

Sue But your entry in *Who's Who* says your hobbies are punning and pondering. Didn't you just say you *don't* like pondering.

Richard Now that I've got more ancient and decrepit, I do wonder what on earth happened in the universe before the Big Bang and things like that. I've got a bit more ponderous!

Sue And do you ponder about other big questions like the meaning of life or what happens when you die?

Richard Oh I think one just snuffs out. And I don't think life has a meaning beyond what we put into it. It's like vision. I mean one not only projects colours onto objects—they're not, of course, themselves coloured—one also projects meaning onto things. If you look at a painting, the viewer is projecting his own meaning into the paint, whatever the artist wants. And ditto with an oak tree; whatever God or Darwin decreed for it, you project meaning into it.

Sue You're being rather coy about your contribution to consciousness studies. I know you have a theory about the function of consciousness. Tell me something about that.

Richard OK, the other big question is what consciousness does. I don't think it's uniquely human. I mean you can stand on a dog's tail and it yelps. It feels it. That's my view anyway. So then you must ask yourself what the function of it is, on the grounds that it wouldn't evolve unless it has a survival function. And what strikes me about consciousness is that it's very much associated with the present moment.

When you're perceiving things, the brain has a vast amount of processing going on from the past. For example, in order to see that cup in front of us I have to have picked up cups in the past, poured coffee into them, probably dropped them and broken them, and done all sorts of things to them. Then I see that cup as a real object, not just because I've got a retinal image and a bunch of signals going into the cortex, but because it's evoking all this from the past. Now it seems to me that you've got to live in the present moment; you've got to survive crossing the road. So it really matters that the traffic light is red or green *now*, at this moment in time, whereas the processes of perceiving are spread out in time. So how do you locate the present moment? I suggest that this is tagged, or flagged if you like, by consciousness. You've got this extraordinary sense of vividness, of qualia, which always applies to the present moment.

Sue So are you saying then that the function of consciousness is to discriminate the past and the future from what's now, and requires action?

Richard Yes, absolutely.

Sue I can think of two objections to that. One is that I can think about the past in the present. In other words, I can bring to consciousness an image of the beach I was lying on on my last summer holiday. How

would you deal with that, because it's a kind of present imagining but it's still a past event?

Richard Yes. But it's very feeble, and it's the *vividness* that signals the *now*. There are interesting exceptions to this, though, and I think one should look at the exceptions. One of them is emotional memories. Let's say you have an emotional memory of shame; suppose you gave an absolutely ghastly lecture and you think back on it, you can sort of blush and think 'Ooh, how can I have done that?'

Sue I do know! I'm glad that happens to you as well!

Richard Absolutely. Now what happens there, I think, is that you get afferent input from blushing, as in the James–Lange theory. You're aware of that input in the present moment, and of course the present moment is always signalled by afferent input. Then this is made special by consciousness, by the qualia.

   Then there's hypnagogic imagery. My hypnagogic imagery, when I'm half asleep, is absolutely vivid as anything—super-saturated colours; and it's partly steerable; it's half conscious, and I can steer and go through these little amazing tropical forests and things.

Sue Do you go flying as well?

Richard Sometimes, yes. But not sounds, I think it's always visual, but very, very, very vivid, no question about that.

   Another exception is vivid dreams, or the effects of LSD, or schizophrenia. With these you can get the sense of immediate reality when in fact it is not the immediate reality, and there I think one just has to say that the system's gone wrong.

Sue My other objection is this: if you say that the function of qualia is to flag up that these things are happening now, so that you can act on them, we know that an enormous amount of immediate action isn't done consciously at all but is carried out by the fast motor system in the ventral stream.

Richard I agree. But that doesn't involve cognitive processing and my theory only applies where there's cognitive processing. Say you've got a simple organism, and it's responding to a stimulus with a reflex or tropism, then there's no problem about the now because its memories and thoughts are not involved; there's immediate action without any problem. But the more cognition you've got, the more there's a problem for the nervous system to separate out the now from the rest of it.

Sue So are you saying that if you look at the course of evolution, consciousness should appear wherever an animal develops in such a way that it faces the problem of distinguishing between the present and everything else it is capable of thinking about? So any animal that faces that problem will be conscious in something like the way we are?

Richard Yes. But we've chosen consciousness and presumably other animals have done the same, but if you were an engineer building a robot you might solve the problem in a different way.

Sue So, conceivably, this might be the kind of quirky thing that you were talking about right in the beginning: where we're all studying the brain science but then some robot-builder comes up with two, or three, or four potential solutions—one of which would be conscious.

But now I'm getting carried away with enthusiasm for your theory, when actually I think it's doomed—that's a bit strong; perhaps I'll change that when I write it down...

Richard I don't mind doomed; it's a good word.

Sue ... because I don't know what qualia are.

Richard Well I do; I know perfectly well what they are. It's only Dan Dennett who doesn't know what qualia are. It bloody well hurts, you know.

Sue I know it bloody well hurts.

Richard So what's the problem?

Sue I think the problem is this: in your sketch of why we have consciousness, it's as though you're saying that consciousness is something added on: here's this machine, doing all this stuff; and then in order to solve a problem it adds on the 'what it's like to feel this,' the 'Ooh it really hurts'. You're implying that dogs might have evolved in a different way, so you could step on a dog's foot and it yelps, but it really doesn't hurt. Your theory is a kind of add-on theory; qualia are something that gets added on.

Richard Absolutely. They are added on in evolution. The earlier mechanism, the immediate-action mechanism, doesn't have consciousness.

Sue But the functionalist would say it's not an add-on; it comes along necessarily with having a nervous system that's capable of yelping.

Richard It runs along with cognition, I think, not with the nervous system, because it's not as sensitive as reflexes, etc. I think it gets into the system when behaviour and perception are heavily dependent upon knowledge—that is, reading the present from the past.

Sue Could you take it out; could you separate it off from the rest of the system; in other words, could you make a zombie? Do you believe in the possibility of the philosopher's zombie?

Richard Absolutely, that then would be like a reflex system or automaton; when you're acting in reflex mode, with rapid behaviour, that's exactly what you are; you become a consciousness-less automaton.

Sue But the classic philosopher's zombie is someone who looks exactly like Richard Gregory, who sits there pondering—perhaps not internally pondering, but saying the kinds of things you do, talking about consciousness in the way you do, drinking your coffee, and apparently enjoying it—and yet all is dark inside. On your theory is that possible?

Richard No. It is for simple behaviour. For simple rapid defensive and attacking behaviour the answer would be yes. But when you've got people thinking about philosophy or having a chat, drinking coffee and all that, then you're using cognition and you've got this problem about making the present separate from the past in your brain. But until then I don't see any need for consciousness.

Sue You've spent your lifetime studying perception, and your wonderful book, *Eye and Brain*, in 1966, really brought to the world the whole idea of perceptions as hypotheses, as guesses about the nature of the world...

Richard And my hypothesis or guess is completely different from the thing I'm guessing or hypothesizing about; the theory of the solar system is totally different from the solar system.

Sue So, in a way, you've accepted the same explanatory gap all along and not worried about it.

Richard I think there's a huge gap, yes. Quite apart from consciousness, there's a huge gap between what a perception is and what the perception is about—what it refers to—sure, but it doesn't bother me. It's the same with a book, for heaven's sake: the description of the Sahara Desert in a book is completely different from the Sahara Desert.

Sue Do you think it still stands, the idea of perceptions as hypotheses?

Richard I do; maybe I'm too stupid to see the objections, but I don't think there are any; I actually think it's right.

Sue I thought it was right too, all the time I've known you, but recently, with the sensorimotor theories which treat perception more as action than as representation, I've begun to wonder whether we need a shift in that respect.

But then maybe the idea of perceptions as hypotheses can survive, because in order to act we must have a hypothesis to act on, but what is being rejected is the idea of a world out there, and a grand representation or mental image in here that is the perception.

Richard I didn't actually define the hypotheses as mental images; they're much more physically based descriptions. Whether the hypothesis has an image or not is another thing—sometimes it does, sometimes it doesn't. But I'd like to say the following thing: that perception is actually amazingly separate from action: if you take ambiguous figures, the perception can flip around from one hypothesis or possibility to another, only one of which has a perceptual consequence. The whole point about it is that the perception isn't tied to behaviour. There's only one motor behaviour but lots of perceptions.

Sue Does your way of thinking about this relate to your work in artificial intelligence? You worked on one of the first ever robots didn't you?

Richard I contributed to it, yes. We started the first department in Europe on artificial intelligence in Edinburgh in 1967, and we did build a robot called Freddie, which was a kind of cognitive robot. But my own contribution was pretty minor. In fact actually my only real contribution was to try to get internal models into the robot. I mean, everybody else was thinking of it as an input-output system and I thought, 'Not on your Nelly. It's got to have an internal model.' I didn't invent that. It comes from Kenneth Craik. So I say it's really got to be a Craikian machine.

Sue And yet now things are swinging the other way, with people doing behaviour-based robotics.

Richard Which I think's rubbish actually. I don't buy that at all, I think it's nonsense.

Sue But in that case we shall see, won't we? It's easy to tell which works better, unlike with consciousness.

Richard So there is a prediction, isn't there, which is good.

Sue I want to go back to something that's still bothering me—this idea of consciousness being an added extra. I keep hitting the same problem, which is the explanatory gap between any kind of brain process and these mystical qualia which you say you know perfectly well what they are, and only Dan Dennett doesn't. I'll join Dennett here and say I haven't a clue what they are. I can sit here and go 'Oh, the brown luscious look of those chocolate biscuits we've been eating'—but I can't capture that; it's always shifting, and I don't know what to do with it. I don't know how it relates to this brain stuff.

Richard I'm not bothered by that; why am I not bothered by that? I don't see a problem at all; why should you be able to capture it? These are sensations generated in one's brain and that's that.

Sue But how can a brain, which is a physical squishy thing with firing neurons, electrical charges, and membrane flows, *generate* the chocolately feel?

Richard Well, that takes us back to Faraday, the magnet and electricity; it doesn't seem to me different, basically, from that, to be honest. Sometimes it's a sign of ignorance that you've got this apparent gap, and therefore it's a goad to find a decent theory; but not to throw away the sensation. That's a given: I bloody well do experience that!

Sue So Dan's gone too far in throwing out qualia?

Richard Exactly; I would say that yes, he's gone too far. What do you think?

Sue I think he's absolutely right. But then I've always liked extreme theories, and I think the only way forward is to throw out all kinds of dualism, because of the classic problem of the interaction between two different kinds of things.

Richard But why does that bother you? It happens all the time.

Sue I think it bothers me in the way that everything about the world bothers me. I wouldn't be a scientist at all if I weren't bothered by things that appear to make quirks in our world; jumps and gaps that don't fit. They seem to me to be an indication of something wrong in the way we're thinking about things; and this seems to be one of those. I think there can't be two separate things: the room we're sitting in, and my experience of the room; somehow the two have to be integrated, and I don't know how ... so I spend my life going around in great intellectual confusion and you don't.

Richard   It doesn't bother me the least bit. I think of my brain formulating internal descriptions of the external world; just as if I look at the books on my shelf, they've got loads of descriptions of things in them.

Sue   And when you think of this self who is doing the describing, who is he?

Richard   It's the sum total of the cognitive processes going on in one's nut, which is separate from the external world, linked only by an afferent signal.

Sue   But isn't that just a sensible scientific description; do you really feel that way?

Richard   Yes, absolutely; I don't see anything wrong with that. I don't think I need to be at one with the universe!

Sue   You're much more down to earth than me. You know, in an intellectual way I would like to be a straightforward identity theorist—that somehow this experience just *is* brain activity; but I cannot for the life of me see how it could possibly be. And I wonder whether, if you were in a brain scanning machine and could actually see your own thoughts as brain activity happening immediately, that the explanatory gap would just disappear—just like we don't need a life force any more, or we can see Venus as the evening star.

Richard   I think that's pushing it too far; it's a good thing to expunge all these different things, as much as one can, but I can't see the point of being terribly worked up about there being more than one thing in the universe; it would seem to me amazing if there weren't more than one thing in the universe—and it would be terribly boring if there were only one thing.

Sue   But we're not talking about things so much as about separate worlds. If you have qualia and physical objects you've got a problem that you don't have if you've got cups and saucers. I have spent so much time staring at carpets—I don't know why it's the carpets that always do it for me; here is this rich red and blue carpet on your floor; you're saying there are qualia, and I'm saying there can't be qualia; help me, what are they?

Richard   Well, if you shut and open your eyes—it's the key experiment, isn't it—that then the qualia disappear and the carpet's still there.

Sue   And how do I know the carpet's still there?

Richard I can still feel it, for one thing. But certainly its existence isn't dependent on the red qualia; it's quite separate.

Sue Ah. You have hit the heart of it here—why do you need something separate from the world of carpets and brains? It bothers me, but why doesn't it bother you?

Richard Well, I think we do need it, because it happens. So we've got to live with it. It's like Everest: it's there, so you might as well climb it.

Sue I give up!

Stuart Hameroff

*Consciousness is
quantum coherence
in the microtubules*

Stuart (b. 1947) originally studied chemistry at the University of Pittsburgh,
and took his medical degree in Philadelphia before training as an anaesthesiologist. In 1973 he moved to Arizona where he has combined his medical
career with a long-standing interest in consciousness, the loss of consciousness under anaesthesia, and quantum physics. He is best known for his
collaboration with Roger Penrose on the theory that consciousness depends
on quantum coherence in microtubules. He is Director of the Center for
Consciousness Studies at the University of Arizona in Tucson.

Sue What's the problem? Why is consciousness such a special and difficult
topic?

Stuart Well, that's the hard problem. The brain is an excellent information processing system, but there's no accounting for how and why we
have subjective experience, emotional feelings, an 'inner life'.

Sue Can you explain how it came to be called that?

Stuart It was at Tucson 1—the first Tucson conference—in 1994. It was
the first ever international interdisciplinary conference on consciousness and we had it all planned out. The first day was philosophy, the
second day was neuroscience, the third day was cognitive science,
and so on.

On the first day a very well known, famous philosopher spoke first and he gave a very boring talk, the second speaker was kind of dull, and so I was getting worried—like the playwright's opening night, you know—that this was gonna flop. Then the third speaker was an unknown young philosopher named David Chalmers, who got up there with hair down to his waist, in a T-shirt and jeans, and gave the best talk I'd ever heard on the topic of consciousness. He talked about the easy problems of consciousness (which include reporting, perception, and things like that), and then the hard problem of conscious experience, which is 'what it's like to be', or qualia, or raw sensations.

After that there was a coffee break and I went out among the people, as one of the organizers of the conference, listening in like a playwright on opening night. And people were just buzzing about Dave's talk and the 'hard problem', as he called it. I think that moment really galvanized an international movement in consciousness, because the problem was identified. From then on we knew what distinguished this field from cognitive science and other fields that deal with how the brain works. They don't attempt to grasp the difficult problem of consciousness itself.

Sue I know what happened there was extraordinary. Dave gave this apparently simple paper, talking about problems that have been around in philosophy for 2000 years, but something about what he said, this label he provided, meant that everyone now talks about the 'hard problem'. What do you think it was about the way he framed it that made this happen?

Stuart Well David would be the first to admit that he was restating things that William James had said, or that Tom Nagel had in his paper 'What is it like to be a bat?' But, as you know, consciousness was under a rock for most of the twentieth century because of the behaviourists, and only came out again in the eighties with Crick and Penrose. I think he just captured the moment; he came along at the right time in the right place, with a very clear message, in plain talk. He characterized the problem of qualia, of why we have an inner life, and he used the zombie example to illustrate it.

This zombie is a hypothetical entity: something like a person but without conscious experience. It might behave like we behave, it might have conversation, it might go to conferences, but it wouldn't have any inner experience or sensation. It's something like a robot or an automaton, or certain science fiction androids. That distinction between a zombie and conscious human was a good way to illustrate the hard problem.

Sue And do you think there could be such a philosopher's zombie, as you describe it, behaving exactly like us, looking like us, saying things like 'I am conscious', and yet for it to be dark inside, for there to be nothing it is like to be that zombie? Do you think such a thing could exist?

Stuart I suspect certain philosophers are zombies!

Seriously, that's not to say Dan Dennett's a zombie, but I do sometimes wonder about Dan, because he tries to explain away the problem of consciousness with a lot of smoke and mirrors. He tries to say that all we are is some form of computation, and everything can be explained on that basis. I just don't think that's true.

Sue So when Dennett says he's got consciousness explained, you don't agree?

Stuart Well, in a joking way I said he may be a zombie. So he himself doesn't have any consciousness, and therefore he thinks he's explaining it. But in all seriousness, in his book *Consciousness Explained*, he really attempts to explain it away. It's a big apology for the AI people. The AI computer industry would like consciousness to be nothing special, and something therefore that could be reproduced in a computer.

Sue But you haven't answered my question. Do you believe it's possible—logically, not that we could make one, but that it's logically possible—for there to be such a thing as a zombie?

Stuart Oh absolutely, because the best computer robot will be a zombie. He, it, she, or whatever, will lack the qualia that we have. It will lack our inner life and our experience. It may have some sort of vision but it won't have *conscious experience* of that vision.

Sue Here you're getting at exactly the problem. You say that you can imagine a creature, with and without this special something—this subjective experience, but are you saying that this is just something magic that we conscious humans happen to have, and the robot won't have? Where does this subjectivity come from?

Stuart Well that's the hard problem.

Sue Right! OK, we've framed the hard problem now. But I want to push you a bit. If you believe in the possibility of zombies, then you've got to have some account of what this extra thing is. Dennett would say there isn't an extra thing; that once you've built some robot that can do everything we could, that's it, there's nothing more to add. You are saying there is something to add, I want to know what that is.

Stuart   You want to know what my answer to the hard problem is?

Sue   Yes.

Stuart   OK, I think there are basically two types of explanation for qualia, for conscious experience. One is emergence; that is, that the brain does a lot of complex information processing and out of that complexity a new property emerges at some higher level. Gerald Edelman speaks about that, and Alwyn Scott has written about it very elegantly in terms of the hierarchical arrangement of the brain, and how novel properties commonly emerge at a higher level in a hierarchical system. One example is the Great Red Spot of Jupiter, or the property of wetness in water. These are properties that emerge from a higher order. However, none of these are conscious, and I question emergence. I think we need something else.

Sue   But, if you believed that consciousness emerges from those complicated processes, then you'd have to agree with Dennett. If a robot or a zombie was actually doing all the complicated things we do, the emergence would happen, and it would have experience just like us.

Stuart   Exactly. So I don't think that view's right.

The other way of looking at it is that consciousness, or perhaps something proto-conscious, is fundamental to the universe; it's part of our reality, much like spin, or mass, or charge. I mean there are certain irreducible things in physics that you just have to say 'they're there' and consciousness is like that. This is the view that Dave Chalmers took in his book, which followed the talk I mentioned. He said that consciousness must involve something fundamental, something that's intrinsic to the universe, and I agree with that.

Now, where we disagree is that he thinks that this fundamental entity, whatever it is, can be attained at various levels, whereas Roger Penrose and I think that the qualia, if they are fundamental, must exist at the fundamental level of the universe, the lowest level of reality that exists. In modern physics that's best described at the Planck scale, the level at which space-time geometry is no longer smooth but quantized. When you go down in scale to roughly 10^{-33} cm you get to this level of space–time where there is a granularity, and that's the fundamental level. It is at that level where we think qualia are embedded as patterns in this fundamental granularity of space–time geometry that makes up the universe. Roger had also suggested that Platonic values in mathematics as well as ethics and aesthetics were embedded there.

Sue But I don't see how talking about the Planck scale, and other levels of physics, relates in any way to the problem we're talking about. That is, that sitting here I'm experiencing a world. There is this complicated world appearing around me, with you and me, my body and yours in this space here. What has that to do with all these microscopic details?

Stuart Your complicated world is described by two sets of laws— Newton's laws and so forth at the macroscopic level, but the bizarre laws of quantum mechanics at small scales. Particles may exist in multiple places simultaneously—superpositions—be interconnected over distances, and time is reversible. The problem is we don't know how small is small. The boundary between the quantum world and the everyday world—quantum state reduction, or the so-called collapse of the wave function—is a big question in physics and seems to have something to do with consciousness.

The point is that our perceived reality—the everyday classical world—precipitates from the 'microscopic details', as you put it, conscious moment by conscious moment. Quantum computers do this—multiple possibilities reduce or collapse to the answer. So in our unconscious minds we have superpositions of multiple possible choices or perceptions which reduce or collapse to one particular choice or perception, say, 40 times per second. Each reduction chooses a set of qualia.

So I would say that the image you have in your brain right now of looking at me, trying to understand what I'm saying, the surroundings, and our environment, is like a painting (if you will allow me a metaphor) and the qualia, the proto-conscious qualia that I'm talking about, are like the paints on a palette. The artist doing a painting has a palette with all these different, simple, primitive colours, and he or she integrates them into a complex scene. So, similarly, I would argue, our brains are able to access the qualia at this fundamental level, but only a particular type of quantum process is able to do that.

Sue So can you explain briefly what kind of quantum process you're talking about, and where it happens in the brain?

Stuart Roger Penrose developed this idea in his book *The Emperor's New Mind* in 1989. He argued, using Gödel's theorem, that our minds do things that are non-computable; that are non-algorithmic. They are inherently different from conventional classical computers. Roger deduced this non-computable element much like Sherlock Holmes followed clues to find the murderer, sometimes very obscure and subtle clues, to find that the only source in the universe for this non-computable influence is the particular type of collapse of the wave function due to quantum gravity at the fundamental Planck scale. Not only does it connect to qualia, it brings in a non-algorithmic—a non-computable—factor which distinguishes our choices from those of computers. So he was proposing a certain type of quantum computing in the brain.

But Roger didn't have a good candidate for quantum computing in the brain, only suggesting the possibility of superpositions of nerves both firing and not firing. I had been studying the computational capabilities of protein structures called microtubules which make up the internal scaffolding within nerve cells. It seemed that microtubules were excellent candidates for quantum computation, that quantum computing might be happening inside nerve cells where they could be isolated. I also knew from my study of anaesthesia that the molecular mechanisms by which anaesthetic gas molecules erase consciousness involve only quantum mechanical interactions with certain proteins in the brain. So it was reasonable to believe that consciousness involved quantum processes and that microtubules might be quantum computers.

It could work like this. Let's say you're looking at the menu at the Mexican restaurant for lunch and you consider the tostada, or the burrito, or the chimichanga. In your subconscious mind you have a superposition of all three of these. Then it collapses and you choose the chimichanga. Maybe some non-computable Platonic value influenced your choice. That's the way to look at volition.

Sue   **It sounds as though you believe in free will?**

Stuart   I have no choice but to believe in free will!

Free will, of course, is one of those very difficult issues, but I think in this approach we can actually explain it in the following way. In the model Roger and I have developed, we have quantum computation in the microtubules inside neurons that reaches the threshold for collapse 40 times a second, to coincide with the 40 Hz gamma oscillations that exist in the brain. And the outcome of each reduction is a process of quantum superposition, quantum computation, which follows the Schrödinger equation, which is basically deterministic.

However, at the instant of collapse there's another influence that enters. This is Roger's non-computable influence which is due to the fine grain in space–time geometry. This has a little influence on the choices, so that choices result from both the deterministic quantum computation and this non-computable influence. The experience of that is free will.

Now I think of it this way. To make an analogy, imagine you've trained a zombie robot to sail a sailboat across a lake, and there's three ports on the other side, A, B, or C, and the wind is shifting constantly. So the wind in this case is going to play the role of the non-computable influences, and the tacking and jibing of the boat are going to be the algorithmic deterministic processes that the robot zombie has been trained to do. But each time he or she tacks it's going to be influenced by this non-computable influence, so that the outcome—the port A, B, or C at which the boat lands—will be a result of both. I think the experience of exerting this deterministic process along with this non-computable influence is what we call free will. Therefore, we occasionally do things that are more or less unexpected even to ourselves.

Sue **You and Penrose have been criticized by people who say that you're just mystery-mongering; you're taking the mystery of quantum physics and the mystery of consciousness and claiming that you can explain one by the other. Pat Churchland once said that 'pixie dust in the synapses is about as explanatorily powerful as quantum coherence in the microtubules.' What do you say to critics like that, who so roundly dispose of everything you have to say?**

Stuart Well, in Pat's case, methinks the lady doth protesteth too much, because she has no explanatory power whatsoever, not to mention the fact she doesn't understand what we're saying. Pat just says consciousness is synaptic computation and ridicules any other possibility. Her view of chemical synapses carrying consciousness is exactly what she said, pixie dust in the synapses. Why should neurotransmitter chemicals cause conscious experience? Actually psychoactive neurotransmitters like serotonin and the psychedelic drugs have high energy quantum states they impart to their receptors and the microtubules inside the neurons. I think altered states occur when we shift more into the quantum subconscious phase. Dreams are quantum information.

If you say there's something other than computation involved in consciousness, Pat and Dan Dennett and others deride it as magic and call you a vitalist. As you know, in the nineteenth century some scientists believed there was a mysterious life force associated with living systems. But as molecular biology became understood the apparent need for an *élan vital*, or life force, seemed to disappear and vitalists were vilified. But the unity and internal communication in living cells remains unexplained, and recent evidence suggests that quantum coherence and entanglement may be an essential feature of life. So call me a quantum vitalist.

But seriously, the position taken by functionalists generates no testable predictions. There's no proposed threshold for emergence of consciousness. What they are saying—that consciousness is a particular property of computation—is not falsifiable, and therefore not really a theory at all. On the other hand what we are saying is testable and falsifiable. Pat and Dan and others can neither prove nor disprove what they're saying, so all they can do is attack what we're saying, usually with out-of-hand dismissals. We could be disproved tomorrow, so we at least have a real theory. You may not like it, but it is a theory of consciousness.

Pat also said that our theory was no better supported than one in a gazillion caterpillar-with-hookah hypotheses, to which we answered—it's not that we're in Wonderland but perhaps their heads are in the sand. And you might recall the *Journal of Consciousness Studies* had a great cartoon about that.

It's not that we're in Wonderland,

But p'raps their heads are in the sand

This is Roger Penrose and Stuart Hameroff's reply to Rick Grush and Pat Churchland who had called their thesis "no better supported than any one of a gazillion caterpillar-with-hookah hypotheses". Rick and Pat are, of course, the ostriches, with Stuart the caterpillar and Roger the rabbit.

Sue You're now well known for your theory of consciousness, but how did you ever get into this tricky subject in the first place?

Stuart In the early 1970s I was at medical school, was interested in the brain–mind problem, and thinking seriously about becoming a psychiatrist or a neurologist. But then I did a summer elective in a cancer lab and, under the microscope, I saw microtubules pulling chromosomes apart in dividing cells. I became fascinated, and even obsessed, by how these little devices seemed to know where to go, and what to do—what their intelligence was, and what was running the show at this cytoplasmic level. Then, because of a breakthrough in electron microscopy, it appeared that the neurons of the brain were full of these same microtubules which had the somewhat magical power of organisation and information processing, and I began to think they were little computers and that consciousness must go all the way down inside the neuron to the level of the microtubules.

Sue Do you now think that your own theories of consciousness have changed your consciousness? Does your theory make you live your life in a different way or feel it in a different way?

Stuart My work has also allowed me to see a lot of the world and meet some wonderful people! I mean I've been doing this microtubule stuff for almost 30 years, so of course it's a big part of my life, but its not my life. I don't rely on this research for my living. Otherwise I probably wouldn't have been able to do it, because it's still an unpopular theory and it would be hard to get funding. So the fact is, that I can have academic freedom because I earn my keep as an anaesthesiologist at the University Medical Centre.

Sue But with that academic freedom don't you agonize about the problem? I mean I walk around a lot of my life, just being there with the hard problem. I'm always thinking—what is this? Why is it like this? That's part of my way of trying to understand consciousness, and the whole process has changed the way I feel about it. Has nothing like that happened to you?

Stuart Well I accept the fact that I am connected to the universe and try to enjoy the interplay between the material world and the enlightened uncertainty of the quantum world. I became interested in the mystical Kabbalah which describes a world of materialistic strife and chaos, and another world of wisdom and enlightenment. According to the Kabbalah, consciousness 'dances on the edge' between the two worlds. I think this is exactly what is happening, consciousness 'dances on the edge between the quantum world and the classical world'. And the more we are influenced and in touch with the quantum subconscious world of enlightenment, the happier we can be.

And I see it every day with my patients, in surgery. In fact that's one of the things that attracted me to anaesthesiology. Every day I put patients to sleep and wake them up and it's still incredible. You wonder—where do they go? And then you wonder where were they in the first place if they would have consciousness?

Sue   Then what do you think happens to consciousness after death?

Stuart   When the quantum coherence in the microtubules is lost, as in cardiac arrest, or death, the Planck scale quantum information in our heads dissipates, leaks out, to the Planck scale in the universe as a whole. The quantum information which had comprised our conscious and subconscious minds during life doesn't completely dissipate, but hangs together because of quantum entanglement. Because it stays in quantum superposition and doesn't undergo quantum state reduction or collapse, it's more like our subconscious mind, like our dreams. And because the universe at the Planck scale is non-local, it exists holographically, indefinitely.

Is this the soul? Why not.

Christof Koch

Why does pain **hurt?**

Christof was born in Kansas (1956) but grew up in Holland, Germany, Canada, and Morocco. He studied physics and philosophy in Tübingen, Germany, and gained his PhD there in 1982. After four years at MIT he moved to Caltech where he is Professor of Computation and Neural Systems and head of the Koch Lab. For many years he collaborated with Francis Crick, searching for the neurological seat of consciousness, and ultimately developing a framework for understanding how consciousness arises from the interactions of neurons in the cortex and thalamus. He is a keen mountaineer and runner. He is author of a textbook *Biophysics of Computation* (1999) and *The Quest for Consciousness* (2003).

Sue   What's **the problem? Why do you think consciousness is so interesting and controversial?**

Christof   Well, the problem is to explain why sometimes I see something and sometimes I don't. For example, there are many illusions vision psychologists have where, just like a magician, you can look at something, and sometimes you see it—sometimes you don't. A related illusion is the Necker cube. When you look at this drawing you can see it in two possible orientations, and your experience tends to flip from one view to the other.

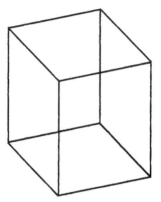

This ambiguous Necker cube can be seen in two different orientations.

So it's a very simple question—where is the difference in your brain? Sometimes you see it one way, you are conscious of it in that orientation, you can talk to your neighbour about it; sometimes you're conscious of it the other way, yet physically the image is exactly the same. Where's the difference in your brain? That is the question.

Sue   **Right, you hit immediately one of the big questions about the neural correlates of consciousness, which is—are we really talking about consciousness?**

Let's say you look at this illusion and the cube flips from one orientation to the other. Now you're able to report that change, and say 'Now I see it this way, now I see it the other way.' Is that really the same as consciousness, or is it only the ability to say those words?

Christof   Yes it is consciousness. Because I don't have to say anything. When you saw it, didn't you have this experience without telling anybody? You didn't say a single word, yet you still had this experience. So I think the verbal part is totally incidental. That's just how you confirm to me that you're seeing it, but you could have just nodded your head. People often nod their heads, or they say 'aargh, aargh, mmm, mmm, oh,' and I don't really need anything much more detailed than that to know they have seen it.

Sue   **But couldn't I be deluded? Some people say that we're deeply deluded about the nature of consciousness, and I could be deluded about this. All I can say for sure is the words that come out of my mouth.**

Christof   No, no, no. There's much more. I have a feeling. I have the feeling that sometimes I see the cube one way and sometimes the other.

I could take the solipsistic point of view—I don't care about you, I don't care about anybody else, I don't care about language—here I am, I'm the only person in the universe, I have these feelings, and then sometimes I don't have the feelings. They wax and wane on a certain timescale. What I want to know is where's the difference in the brain when I have those feelings.

Could I be deluding myself? Yes, in principle. But unless there is compelling, empirical evidence to the contrary I'm going to assume I'm not because these experiences are such a salient part of my life. I'm going to assume they're real, and I'm going to try to track down the neural correlates. Then, once I have the neural correlates, everything is much more concrete because now I can say—OK, if these neurons are synchronized in this part of the brain, now let's artificially get those neurons and synchronize them. Then, I put it to you, you will have the feeling. Now that's a testable proposition. As long as I can manipulate it, and I can move from correlation to causation, I'm happy.

Sue Can you give me a simple example of how you can move from correlation to cause?

Christof An example is another illusion. With epileptic patients, neurosurgeons need to discover where the seizures originate by inserting electrodes into their brains, into a high level area, the medial temporal cortex. We can then listen to individual neurons. We use an illusion called 'flash suppression' where there are two stimuli present, and sometimes you see one and sometimes you don't, even though both pictures are always physically present. Now, suppose we have a cell that only responds to an image of a car—and we've many cells like that. When you see the car the neuron fires; when you don't see the car—because it's suppressed by another percept, even though the car is still physically present on your retina—then the neuron doesn't fire. I can now do another experiment and ask you to close your eyes and think of a car, and the neuron fires. Then I might have just shown you a picture of Bill Clinton and the neuron doesn't fire, whether you see him or imagine seeing him. So this neuron fires when you are thinking of the car or seeing it, but doesn't fire when the image is physically there but you're not seeing it.

That's a very tight relationship to consciousness. It's a correlation, not a cause, but in principle you could go inside patients' brains with little electrodes and inject tiny currents that don't cause any damage, and stimulate those neurons. That way I can try to move from just correlation to causes and that's certainly not out of the question. So that's pretty cool, you've got to admit.

Sue   **If you stimulated those neurons and the patient said they had an impression of a car, is that the end of the story for you?**

Christof   No, no, no. Because then you want to know where the neurons are projecting to. What happens if I take those neurons and I inactivate their targets? Let's say that the neurons I'm stimulating all project to the prefrontal cortex. Now let's go to prefrontal and see, if I eliminate the target area but I leave the original neurons still intact, will the person still have a percept? Which of the targets will be really essential and which non-essential? I want to walk through the entire brain and really characterize much more about the NCC. Is it always the same neurons? When you have a percept of a cow, or of Bill Clinton, or something else, is there something special about those neurons? Do they all look the same? Do they all project to the same place? Do they use the same type of neurotransmitter? All those are questions you can ask.

Sue   **Well let's imagine, and this is probably not that far off, this wonderful time where all that is possible. Now we can see all the information flowing through the brain, and exactly where it's all going. There's a great temptation to think—right, now we're going to find the *place where consciousness happens*, or the particular neurons in which it happens, or the particular pattern, or group of neurons, or whatever it is that your theory predicts you should look for. Because there's a kind of mystery there—what would make those neurons, or that pattern, suddenly produce subjective experience while the others don't?**

Christof   Inherent in your question is this scepticism—OK so now you've told me it's these neurons that have a standing wave between the inferotemporal cortex and prefrontal cortex, and feedback, and then you're conscious—fine, but why does it give rise to the subjective feeling?

Sue   **Exactly.**

Christof   Right now the answer is 'I don't know'. Let me explain why I'm not too worried about your concern right now, because of this very vivid analogy with vitalism.

There have been convinced materialists, such as the English biologist William Bateson in 1916, who said 'I do not understand it. It's inconceivable to me how the entire specificity that must be inherent in a single cell can be passed on from one generation to the next. I know chemistry; I know this can't be done.' And so, in response to the inability of science at the time to appreciate the existence of highly particular macromolecules, people postulated the *élan vital* and all these other things. What was missing was that they had no idea of the prodigious amount of information that you can store in one molecule. They didn't even have the idea of macromolecules.

Likewise I think we should be very careful.

Sue **But in the case of vitalism it was a question of understanding the process; the information storage, the copying and so on, and all that could be done with objective third person methods. But in the case of consciousness, we have this peculiarity that it's about subjective, first person experience. Now I think you're saying something like this ... let's not worry about that really hard problem, because when we understand all the rest it will disappear. Is that what you're banking on?**

Christof Yes, yeah. The only way we've made progress is by doing the hard science. We're relentlessly trying to push this sort of approach to its limits and then see, in the fullness of time, whether we'll be able to explain everything. It's possible that there are things that, as Chalmers has argued, are forever beyond us. At this point I've no idea.

But as a scientist I can say the following: I can go back to Plato, or to Descartes, and for the past 2,300 years we've not made any progress on the philosophical aspects of consciousness. Philosophers have been profoundly wrong in almost every question under the sun over the last 2000 years. You should never listen to the answers of philosophers, but you should listen to their questions.

Philosophers pose interesting questions, but their answers usually are not very useful or meaningful. Scientists are very different, you tend to be more humble because you know you've a very limited ability to understand a system even with three or four variables. So you know all this knowledge is provisional, and we have to wait and see how it comes out. So therefore I just don't see any reason why I should not continue to do what has been spectacularly successful over the last 200 years.

Sue **Nor do I. But don't you think that any recent philosophers working on consciousness have made any kind of contribution, or any step forward?**

Christof These are all good colleagues of mine. I like them personally; in my life as a scientist I think what philosophers have done for me is help me clarify certain problems. For example, there's the language I use now. When I talk about cause, and whether the NCC are sufficient for consciousness, or cause consciousness, I am much more careful. There's no question that philosophers have made a contribution here, but I don't believe any of these long elaborated arguments that are mainly based on language games and lead them to conclude that consciousness exists, or doesn't exist, or can never be explained. For me the quest for consciousness is primarily an empirical problem. So let's push it very hard and then see.

Sue How did you get into studying consciousness in the first place?

Christof Well, I did a minor in Philosophy! I was in Tübingen with its classical German idealism, and I studied Kant, Nietzsche, Schopenhauer, and all those. But the first time I thought very hard about the problem was about 18 years ago when I was in pain. I had toothache, and I wondered why does it hurt? I couldn't for the life of me figure it out, why does it hurt? The conventional explanation, the medical explanation, was that there's this inflamed tissue in the tooth and it triggers the action potential that travels along the trigeminal nerve, inside the spinal cord, moves up and somehow gives rise to neurons firing in the brain. But so what? That's just ions sloshing around; sodium, potassium, and chloride ions sloshing around. Why should they hurt? And when they move in and out of another cell, they give rise to pleasure or they give rise to the feeling of seeing. Why does it fundamentally hurt? That really was the start of it.

Sue What do you think about pain now? Do you think you're any closer to understanding that magic whatever-it-is that seems to make those neuron firings and those ion exchanges become the 'hurtingness' of pain?

Christof Francis and I believe that we have a better understanding of sort of the framework we need. But if you ask me whether I know why some neurons are involved when you get this feeling, I don't. We have some ideas where meaning comes from, so I can explain why certain things are more meaningful than other things, but I don't understand why some neuronal activity feels like something. I mean I really don't know. But I don't go out and say—well this calls for a fundamental revision of our thinking, or a fundamental new law, or that it can never be solved.

Sue Do you believe in zombies? I mean the possibility of a philosopher's zombie?

Christof No.

And I don't think there are going to be these NCC neurons, those ten neurons that if you knock them out you're a zombie. But there's going to be something specific about the neurons that give rise to consciousness, like their specific morphologies and specific projection patterns. And if this is true, of course, it makes our job immensely easier as neuroscientists.

Sue Do you believe you have free will?

Christof Probably not.

Sue How do you cope with that in your life?

Christof If you read Kant (sorry another philosopher!), he argues that we have to act like we have free will because it accords with our subjective experience. I mean nobody forces me to lift my hand. You didn't force me to lift my hand. I did it of my own cold free will. From the legal point of view we assume it exists. We punish people, and I think we should punish people, if they transgress the laws, assuming they have free will. But was it really free will in the metaphysical sense? I think it's a very difficult question.

Free will in the metaphysical sense really implies there's action without any physical precedents. Now as scientists, or even as any thinking person, we know that can't be the case. There always have to be physical precedents. So I only mean I am free in the sense that it's not you who is determining my actions; it's not blind force or destiny; it's my upbringing, and my genes, and my predilections, and my desires. All of this, plus some random component depending on fluctuation and noise in my brain, comes together in making a decision one way or the other way.

It doesn't bother me too much, no.

Sue For some people it does. Some people find that it causes real awkwardness in their life, and for their moral decisions, and so on. Why do you think you don't find it a problem? Is it because you struggled long and hard throughout your life with it and came to a happy stability with the idea? Or do you think you have some argument that makes it not seem so difficult or painful?

Christof It's a good question, I never thought much about it. I know it just doesn't bother me. Maybe ultimately it depends whether you are a control freak or not. You really have so many things that are beyond your influence. There are few things that I can control, where

I think I am in control, where at least I managed to delude myself into believing that I'm in control and they are initiated by me...

Sue **You're talking about 'me'. Are you referring to 'me' as in this physical organism sitting here in the chair with a lovely pink shirt and a purple waistcoat, or are you referring to 'me' as in something inside that lives in the body, looks out through the eyes, and drives it around?**

Christof Subjectively of course I'm referring to the latter one. There's a Christof sitting inside me. I can tell you exactly where he sits. It's exactly here, between my eyes. If I'm blind I assume I would pin-point it somewhat differently because, like most people, I think I'm between my eyes because I'm a binocularly driven creature, and that's my personal experience of the me looking out at the world.

Now I know perfectly well, from a neurobiological point of view, as Thomas Metzinger and other philosophers argue, that there is no true me—or the me is subject to ever changing fluxes, and the me today is not quite the same me as yesterday or as the me looking at the picture ten years ago. But from a subjective point of view it's a perfectly coherent concept that there is a Christof sitting inside my head, and looking out at the world.

You may think of it as a very compelling illusion, but from a personal, subjective, phenomenological point of view I'm quite happy with it.

Sue **This is very interesting, because when it comes to free will or the sense of self, you're happy to say 'I know these things are illusions, and I'm happy to live with them' but when it comes to the idea of things being *in consciousness* or *out of consciousness*, or the concept of you being conscious of something at a given time, and not a moment later, you're not prepared to say that that is also an illusion.**

Christof OK. So Dennett may be right. I do not have a rigorous proof—I'm not sure you can have a proof—but I have experiences of the world and those are the corner stones of what I know. I can apply radical scepticism, but ultimately I feel some things and I don't feel other things.

Right now I don't feel the state of my stomach; I've no access to the pH in my stomach. Now there are roughly 50–100 million neurons down there, in your gut, called the enteric nervous system. It's a very complicated, sophisticated nervous system; they're doing all sorts of things you don't want to know about. Why don't I have any feelings there? Likewise, I don't know the state of my immune system, because

I don't have any conscious access to it. But I do have conscious access to certain parts of my brain, and for me there's a fundamental difference between the two.

I had this correspondence with Dan Dennett about this. I'm a climber and hiker, and I was in the Sierras and I had to abort my climbing trip because I had a bad toothache—once again the toothache. He wrote me a letter saying that the Crick–Koch programme was sort of delusional; there wasn't any NCC because there wasn't fundamentally any real conscious sensation. I had just got back and I told him that I had to abort this climbing trip because I had this very, very awful feeling. It's not just about the behavioural disposition that I go and rub my mouth and I moan and I say 'Ooh, ooh this is so bad.' I really had this bad evil feeling in my head. And you can't just say you're linguistically confused because at that point the pain is the most annoying thing that there is. Right now, you really have a bad toothache and you don't have pain medication because you're out there in the mountains. It's not very convincing to tell me 'Sorry, you're just linguistically confused.' It just doesn't cut it.

Sue You said you don't have conscious access to all this stuff in the enteric system and immune system. What do you think you mean by 'conscious access to'?

Christof That I have no neural representations in my brain, or in my body in general, that represent this information in an explicit way and make it accessible to the planning stages of the brain.

Sue So it's the planning stages of the brain that are critical here?

Christof Yes, because that's what Francis and I think is the function of consciousness—to make a summary of everything around me that's currently relevant, and to send that summary over to the planning stages to make the next decision about what I'm going to do next.

That's why we have a theory of zombie systems, in the non-philosophical sense. These are automatic systems that control my eye movements, my enteric nervous system, that allow me to run and climb and drive and do all those things. They all do very complicated things but they bypass consciousness. You don't need any consciousness for stereotypical things like that, but if there's a funny noise, or there's an earthquake here, then you will really have to think—where are we? Where do I get out? That's what I need consciousness for.

Sue So in your view, consciousness *itself* has a function?

Christof Or, if you want, the neural correlates of consciousness have a function. I don't believe in Ned Block's distinction between *P* consciousness and *A* consciousness.

Sue You think they're the same thing?

Christof Ned has never given us a clear empirical or operational way to distinguish them. It may be possible that conceptually they're different, but as long as I can't operationally distinguish the two, I'm not going to worry about it.

Sue So when you talk about 'the function of consciousness', you actually mean 'the function of consciousness-and-its-neural-correlates' or what's happening in the brain?

Christof Yes.

Sue How has studying all this changed you as a person?

Christof Well, I can tell you in a very practical way, I don't squish bugs anymore. I'm very serious—unless they attack me. Why? Because I'm a biologist. Most pet owners would agree that cats and dogs are conscious, and the monkey is conscious. Monkeys don't have the same richness of consciousness as you and I do; they don't know about death, and Macintosh, and representative democracy, but they feel and see, and their brain is very similar to ours.

Now, you can ask, how low does it go in the evolutionary ladder? What about bees, for example? It's amazing how quickly you can train bees to do very complicated pattern recognition, including tasks that require the online storage of information for tens of seconds. In humans this always requires consciousness. At least, whenever you have a patient that has impaired consciousness, he or she can't do those tasks that require this sort of short term storage of information. So then you'll realize that you're not really sure anymore to what extent these bees are conscious. This raises the question—what's the minimal nervous system you need? Do you really need 20 billion neurons?

Sue How many do bees have?

Christof A million roughly, give or take, compared with our 20 or 50 billion.

Sue And a completely different organization from us?

Christof Yes. Their neurons are similar to ours; you can record action potentials; they have synapses; they aren't fundamentally different. But they don't have a cortex, and they don't have a thalamus. So the internal structure's quite different. But they do have feedback pathways, and they have recurrent networks, so I don't see in principle why you can't get the same representations or similar representations you have with mammalian cortex. I'm not saying every animal is conscious. Take the roundworm, *c. elegans*, for example. I'm not sure it has enough sophisticated behaviours for consciousness. The operational way to test this is to ask whether any of these critters have sophisticated behaviours that are non-stereotypical and not inborn, and whether you can relatively quickly train them to do new things. With bees you can do that. It's amazing what some scientists have done in terms of long distance homing, pattern discrimination, and so on. I'm really not sure anymore that these creatures aren't sentient in some way. Then what right do I have to kill them, if they're not just automata but can actually sense and feel.

Sue Do you eat meat?

Christof (sigh) Yes.

Sue It's a hard one isn't it?

Christof Yes. I try to eat less meat but it just tastes so good.

Sue I infer from this that you don't think much of those theories that tie consciousness to language and say that no creature without language could be conscious.

Christof No. I've never seen any convincing evidence to show that without language I wouldn't see red or feel pain.

Sue You've just described the reasons why you think a large range of animals may be conscious, but then you ask for evidence. How could you ever find out whether any animal is conscious or not?

Christof OK, well, how can I know you're conscious?

Sue I'm not.

Christof OK, well most people would assume that they are conscious and I assume I'm conscious. I also assume you're conscious! Why? By analogy, because your brain is statistically speaking indistinguishable from mine, your evolutionary history is the same as mine. If I step on your toe, you will behave roughly as I would if you stepped on my toes.

Now the monkey is a little bit hairier than me, looks a little bit different, doesn't talk, but has a brain that's similar to mine, has an evolutionary history that we shared except for the last 13 million years. If I get the monkey to do a visual experiment it behaves very similar to your typical undergraduate subject. If I take a little cubic millimetre of monkey brain, very few people on the planet could distinguish it from a little bit of human cortex, without using very elaborate tools. So by analogy I would say it's probably also conscious.

So all you have is an analogy and it becomes more and more difficult as you get more evolutionarily distant. Ultimately we need a theory of consciousness that's not just based on similarity with humans, but that tells us which systems have subjective states, which artificial systems have subjective states. What about machine consciousness? What about the Internet? Ultimately a complete science of consciousness would have to include such a theory.

Sue And then we would be able to say which things are conscious and which are not. So would I be right in saying that you are happy to wait for that day, and you think that it will come? And so for the moment you're not going to worry about the problem that you can't know for sure whether I'm conscious or not?

Christof Yes. Because as a scientist, right now, I know there's nothing we can do about it.

Stephen LaBerge

*Lucid dreaming
is a metaphor for
enlightenment*

Stephen (b. 1947) originally studied mathematics and chemical physics, before taking a break and then returning to work for a PhD in psychophysiology at Stanford. This included his pioneering work showing that lucid dreams really do take place during REM sleep. Since then he has continued research on lucid dreaming and the psychophysiological correlates of states of consciousness at Stanford. In 1988 he founded the Lucidity Institute. His books include *Lucid Dreaming* (1985) and *Exploring the World of Lucid Dreaming* (1990).

Sue   What is it about consciousness that makes it so interesting?

Stephen   *Consciousness* makes consciousness interesting. It's exactly that self-similar quality, the fractal nature of it, which makes it so endlessly fascinating.

Sue   Tell me how you got interested in it in the first place.

Stephen   I started out as a hard scientist, studying chemical physics at Stanford. I had a very limited view of the world and then in California in the late '60s, I had experiences with psychedelics that suddenly opened me up to the possibility that there was another universe that I hadn't realized was there: the inner world, so to speak.

I learned one important lesson from LSD: under its influence I saw living breathing, hieroglyphics superimposed on a blank wall, and thought, 'Ah, so this is what the world is *really* like, overflowing with meaning, beauty, and complexity. How could I not have seen it before!' But then the next day, 'Ah, wait a minute, *this* is what it's like, *that* was just an illusion.' And finally to realize, no, it's neither like this nor like that, those are just my mind's understanding of what the world is, and the world remains a mystery. But I soon found out that drugs were not useful for more than giving one a glimpse of what the possibilities are.

Then after a long and strange path I found myself returning to Stanford to do research in psychophysiology, and that's where I did work proving lucid dreaming actually happened. It's actually rather surprising to realize that 25 years later I'm still working on this one area. I had no idea how vast a topic it was, how much there was to learn, and how far we have to go.

Sue   **Lucid dreams are dreams in which you know, at the time, that it's a dream.**

Stephen   Exactly. In most dreams we are conscious on an experiential level: for example, a strange thing happens to me, I wake up in bed and I tell a story about being at the circus. The fact that I can remember those experiences means that they were conscious in the sense of the reportability criterion, but what's usually absent from dreaming is the reflective consciousness that everything that's happening there is happening in a dream; that it's all in your mind; that you're in fact asleep in bed. When you remember this you now have a new set of possible actions that make sense in this wider context, and which before were literally unthinkable. It's like saying there's another dimension that I'm in contact with. That sounds kind of crazy, but that's what it's like to be in a dream, in the laboratory, hooked up in the physical world, with these wires on you, talking to a dream character and saying, 'Excuse me a minute, I have to do this experiment.'

Sue   **But it seems to me that when you become lucid you feel as though you've woken up in some way. I can think of the very early lucid dreams I had: in one I was going up a ski lift; it was dawn, and the sun was rising, and I thought, 'It's very strange that the ski lifts have opened so early in the morning.' Then I realized I was nearly at the top and had to get off but I didn't have any skis on. I thought, 'This is terrifying, how am I going to get off the lift without any skis?' And then I thought, 'Well,**

how did I get on it without any skis?' That's when I realized it was a dream, and at that moment everything became vivid and beautiful and clear. With the realization that it was only a dream, it seemed to become more real. What on earth is that about? What does that tell us about consciousness?

Stephen Yes, it does seem paradoxical, doesn't it? Why should realizing something isn't real make it seem more real? I think this enhancement of vividness is due to our intense focus on the present. *Here I am!* Right now, if you were to realize the miracle of consciousness and to be *here, now*, you'd have a similar experience.

It's like that because of the great novelty of being in a dream, and looking around and seeing it as real as, or more real than, what you've been taking for reality all of your life, namely waking existence. This is such an amazing experience to people that they find it exhilarating and that enhances their awareness of that moment.

Sue I imagine that when you started your work the behaviourists wouldn't want to have anything to do with this. Certainly scientists didn't want to study dreams at all for a long time, but even once dream research got under way I don't think lucid dreaming was exactly a popular topic, was it?

Stephen No, it wasn't. It was worse than unpopular; it was impossible at first. Of course, *I* knew that lucid dreams were real because I had experienced them myself, and reflected on them much as you just described. I remembered the condition of my body in bed: it's winter, I have heavy covers over me and I don't feel them, there's a clock next to the bed ticking away but I don't hear it, which means I'm not in sensory contact with the physical world, and therefore in a basic sense of the term, I'm asleep. So all this has to be in a bona fide dream; no question about it. I knew that experientially, in the first person. But how could I prove this to someone else, especially a sceptic who says it's impossible, because, 'How can you be conscious while you're asleep?' Framed that way it sounds paradoxical; but if you frame it as, 'How can you be conscious *of the fact that you're dreaming* while you are unconscious of sensory input from the environment?', there's not so much of a problem.

Sue I bet making that argument didn't convince sceptical scientists.

Stephen No. What was required was evidence. I was aware of the earlier research showing that the direction of gaze reported in dreams sometimes corresponds very precisely with measurable eye movements. So I thought that in a lucid dream I could look to the left and right, left and right, and thus make a unique and easily identifiable signal that would be a symbol meaning 'I now know I'm dreaming'. If I could do that in the laboratory, then we could see, by the physiological data, whether or not I was awake, or in REM sleep, or in some other state or mixture of states. It turned out that signal-verified lucid dreams occurred almost without exception in unequivocal REM sleep, not a partially awake state, but the most intense form called 'phasic'.

Sue   Presumably you had to make these signals with eye movements because all other parts of the body are paralysed?

Stephen   Exactly. The problem was how you could say 'Yes I know I'm dreaming now,' and find out what stage of sleep one was in. In REM sleep most of the body is paralysed, mainly the muscles of vocalization and locomotion, but respiration is not paralysed, so you can actually signal with your respiratory movements as well, but the eye movements are the easiest to use. This pattern of paralysis is presumably due to evolutionary selection pressures. Eye movements never caused dreamers to fall out of trees, and thus happily left open a convenient link to the outside world during REM sleep.

But then, having used eye-movement signals to validate lucid dreams, on our first attempt to get it published in *Science* magazine, one of the reviewers said, 'This is a wonderful breakthrough, a new technique' but the other one said, 'Well, I don't know quite what's wrong with this study, but it can't be true because it's impossible so you must reject the paper,' and that's what happened. So then we submitted it to *Nature*, and *Nature* said, 'of insufficient general interest'. So it took two years to find a journal to publish in, and this was *Perceptual and Motor Skills*, a second-line journal used by sleep and dreams research, but only after more reviews saying, 'Yes, this really is true' and answering a large number of objections. It was very hard for people to accept. Indeed, the significance of these experiments hasn't yet sunk in for most researchers. For example, marking an event during a dream with a lucidity signal also should have set to rest the mistaken idea that dreams are not experiences, but some people still credit Dan Dennett's fanciful 'cassette theory' of dreaming.

Sue   And I'm one of those people! I think ordinary dreams might not count as experiences, but lucid dreams do. I mean, doesn't becoming lucid change everything?

Stephen OK, what happens when you 'become lucid'? Essentially, you become explicitly aware of a particular very important fact—that you are dreaming. You haven't changed everything, just your metacognitive interpretation of what is happening. You've changed how you're thinking about your experience—yes, I do mean experience! You don't think, 'Oh, a moment ago I was "unconsciously composing cassette memories to be loaded later as a so-called dream" but now I'm having real dream experiences.' Now if you were sleeping in a sleep lab with electrodes to record your eye-movements, you could mark the moment when you became lucid by, for example, looking to the left, right, left, and right in the dream. Then let's say you flew about your dream and then woke up a few minutes later and reported your dream. The polygraph would in fact show the eye-movement signal just when you reported. How would the cassette theory explain this? You unconsciously moved your eyes in that particular pattern and when you woke up you somehow miraculously remembered this and wove it into a convincing story!? If you find this account even mildly plausible, just consider a more complex example such as the one you reprinted in your text book. For once, the common sense explanation makes more sense: dreams are experiences.

Sue **One thing that happens to many people when they become lucid is that they feel that they can control things. Have you been able to learn about this with your experiments? Can you control anything? Are there limits to what you can do?**

Stephen The first kind of control you have in dreams is the same as you have right now. For example, we might change the tape, or move over there, but a lucid dream is not a state where everything you're doing is deliberate. It's rather a state in which you have more choice, in which there are more possibilities open to you, because now you understand that there is another world, and another life, and what you're doing in the dream may have very different meaning.

As to what people usually mean by dream control—that's something more like 'magical' dream control. The Tibetan Buddhists who have been practising the yoga of the dream state for 1000 years claim that you can change dream content in any imaginable way: that if it's single you can make it multiple, if it's hot you can make it cold, small, large, and so on. They believe that it's possible to change it all in any way you like.

Now, in terms of the actual experiments we've done, most of our laboratory experiments have focused on some kind of simple activity of which measurements could be made to determine how closely the dream state corresponds with doing a similar task while awake. For example, we've measured dream time by having people make an eye-movement signal, then estimate ten seconds by counting 1001, 1002, and so on, marking the end with a second eye-movement signal. We then measured on the polygraph record how long that took, and compared that to the waking state and found that essentially the sleeping and waking times were the same. We did a number of experiments like that with simple actions, but I haven't personally experimented to see whether I really can do absolutely anything. I've been more interested in finding ways to respond flexibly to whatever comes up in the dream, because what I'm interested in is developing my adaptiveness to life in whatever form I find myself in, whether it's awake, or in a dream world, or in some other world.

Sue So you've got two things going on at once here: one is the scientific research to find out objective facts about first person experience in dreams, and the other is how this research is affecting you. You're actually changing your life through doing this research, aren't you?

Stephen Yes, it's really a matter of personal exploration, of addressing questions like: Who am I? What is it to be a being? What is it to be embodied in the world? The body that we experience right now, the thing we call the physical body, is really the phenomenal body, or body image. Now, in a dream you also experience the body image. Yet you say, 'But isn't that one just dream stuff?' That's what *this* is right here, and if one takes seriously the insights that one experiences in lucid dreaming, it can profoundly change the way one looks at the world. I really do believe that what I'm experiencing right now, while we're having this conversation, is a kind of dream; a special case of dreaming: dreaming in which what I am dreaming is constrained by the sensory input from whatever that thing called the physical world is. That's how you and I can share dreams together: my experience is in my mind, and yours is in your mind, and we happen to be interacting in this third space called the physical world. But oddly enough, we don't really know how those different spaces relate to each other. We don't know whether it really makes sense to think of a mental space entirely separate from the physical space, or whether they are in some sense the same thing.

Sue Isn't this getting close to the central problem of consciousness? How can it be that there are different kinds of thing in the world? I can say that there's something it's like to be me; I can *believe* that there's something it's like to be you and because we seem able to talk about things and agree about them we think there's a third world—a physical world—which is somehow different from these two. What do we do with that problem?

Stephen I'm not certain that it's so much a problem, this third world, this idea of physical reality: it's more like a hypothesis. It's how we explain the correspondence between your mental experience and mine. In the dream state I could do a reality check: I look at my watch, I look away, I look back. It's a digital watch, so it would be very likely to change if this were a dream. But this is the third time I've looked back and it still hasn't changed, so I say 'I must be awake'.

If I were dreaming instead, I'd be talking to a dream figure who wasn't actually there. But the most interesting part is not when I focus on you, the other in the dream, but when I say, 'What about me?' I say, yes you're a dream character, and yes this is a dream table, I'm sitting in a dream chair, and this must be a dream shirt I'm wearing, that must have been a dream watch, this is a dream hand, and *this* must be a dream Stephen!

And then I realize that's me, that's who I am, and given what I said earlier about there being an exact equivalence between the body we experience while awake and the one experienced while dreaming, then you have to realize that what I always thought I really was, is just a dream; an idea. And then one finds that one really doesn't even know what reality is.

Sue I've often heard it said that if people question whether it's a dream or not they pinch themselves, but I've tried that, and it doesn't prove anything, does it? You feel the dream pinch. That is, unless sometimes it comes a bit late, or there's something else funny about it.

Stephen Well, we asked our lucid dreamers to carry out an experiment to compare three different kinds of sensations across three states. We asked them to pinch themselves, to caress the forearm, and to press the thumb while awake, and then to rate the intensity of the sensations, and their pleasantness or unpleasantness. Then we asked them to do that in their imagination as well, and then in a lucid dream. Then we compared how these intensities varied, as well as the pleasantness/unpleasantness dimension.

Sue How could they tell you the answers in the dream?

Stephen Not in the dream, but after awakening! The basic results were that the pressure sensation was quite similar between both the waking state and the dream state; with imagination it's much less, as you can imagine. For the caress sensation, the pleasantness was higher in the dream state than the waking state. That makes sense because it's not all that pleasurable to gently stroke your forearm, but in the dream it's a more curious mixture of things—maybe it's like a schizophrenic tickling himself. But the biggest difference was the pinch, which was much less likely to produce pain in the dream than in the waking state.

I did this myself; I was surprised when I pinched my skin, it felt like rubber, but there was just no pain. I wanted to find out why, so I took a pencil and stabbed my hand and owwww: yes I can feel pain in dreams, but it's not a reliable sensation, it's not guaranteed to happen. This may be because REM sleep is more likely to activate the reward areas of the brain, than the punishment areas.

Sue Why is it so difficult to become lucid in a dream, and so difficult once you've become lucid to stay lucid? I often wake up from a really bizarre dream, in which completely impossible and ridiculous things have happened—and I think, 'Why didn't I realize it was a dream?'

Stephen The usual answer is that there's something defective about our minds: there's a failure of higher cognitive function in the dream state—the assumption being that similar bizarre changes would be immediately noticed in the waking state. Of course, recent research on change blindness tells us otherwise. So when a dream character suddenly changes into 'someone else', low-level change detectors cannot compare sensory input to working memory because the system is functioning in the absence of sensory input. The fact that we *do* sometimes notice and properly interpret anomalies as dream signs shows that higher-order metacognition can be fully compatible with REM sleep. So it's difficult to become lucid for the same reason it's difficult in the waking state to notice anything we're not attending to. Novice lucid dreamers tend to lose their lucidity because they become emotionally involved in the dream events and lose the broader perspective. But that tendency can be overcome with a bit of practice.

Incidentally, we've just finished an experiment with Luis Buñuel's film, *That Obscure Object of Desire*, that is *a propos*. Only 25% of some 150 viewers noticed that the central character was played by two different actresses in alternating scenes throughout the movie! Does that sound like what the waking state is supposed to be like? That's the problem, it's *not* like what people think; and few dream theoreticians take the trouble to do comparisons with the way our consciousness actually works while we're awake.

Sue One implication of change blindness is that the apparent richness and continuity of our visual world in ordinary waking consciousness is a big illusion. Are you saying that your own research suggests that both waking and sleeping are similar illusions, rather than waking perception being the real thing, and dreaming being deficient?

Stephen Yes, I think that both states are really the same brain in two different conditions, trying to do the same thing, namely to understand what's going on around me so I can get what I want and avoid what I don't want. So the world is an illusion in the same sense that everything you see on television is an illusion. It can be either manufactured with computers or on the stage or a newsreel so you can't tell from the fact that it's an illusion whether it's truth or not; and the same thing applies to the world. Yes, the world is an illusion, but, as some mystical traditions claim, the truth is always being shown there.

Sue This implies that if we have these two related kinds of illusion and you can wake up in a dream and say, 'Oh, but now I realize it's a dream,' you might be able to wake up in waking life in the same way—and have lucid living.

Stephen Yes, certainly. The religious, esoteric, religious traditions of enlightenment talk about that exactly, and lucid dreaming seems to be one of the best metaphors for what that enlightenment would be like.

Here you are in a dream that you don't know is a dream, and so you have a very limited view of what your possibilities are, who you are, what you're doing there, and what really matters. Suddenly you remember that you're dreaming and that changes everything. And in the same sense with enlightenment, it's said that one comes to understand a deeper level of unity. Normally we are acutely, uncomfortably aware of separateness and the fact that there's a great distinction between Sue and Stephen. You're over there and I'm over here; but there's another level on which we both have something in common: not the self, but the 'I', the experiencer. When you tease this apart you find out that there's no way to distinguish the ultimate nature of that experiencer in Stephen or in Sue, because the stuff that distinguishes—Stephen's name, his birth-date, all his physical characteristics and all that—is the stuff which is not necessary to being who I am.

Sue You're saying that if you were to wake up in waking life, which might be called enlightenment, somehow this separateness would disappear; the self would disappear? Yet in a lucid dream it almost seems the other way round: when you wake up you feel *more* yourself, as though before I became lucid it wasn't really me dreaming, but now I'm actually here in my dream.

Stephen Yes, but it depends on what you mean by 'yourself'. Do you mean, 'I feel more like who I am,' or is it this person that people call Sue Blackmore? You don't feel more like the outside view of you, you feel more like the inside you, and that's the point: to really feel that identity is something like the difference between snowflakes. Suppose we take ourselves to be individual snowflakes with a particular crystalline form. Certainly there's a difference between the two, they have different structures. And here one snowflake is falling into the ocean; what does it fear? 'I'm about to be annihilated, I'll disappear, I'll be gone, nothing.' But perhaps what happens instead—and this is a metaphor for death or enlightenment—is an infinite expansion, as you remember that you're not just that one drop of frozen water, but that you are *water*. So this metaphor of substance is another level that is simultaneously present with the form; the separation doesn't disappear: it's just that it's only the form; the substance is unity.

Sue You're not only challenging ordinary science by talking about something as dodgy as lucid dreaming, you're really going the whole hog here in talking about mysticism and self-transformation. This isn't normally part of science, is it? Do you think that a science of consciousness must necessarily entail these questions of self-transformation?

Stephen Yes. There are many kinds of knowledge which we have to distinguish. Certainly scientific knowledge is exceedingly important, and if I can have scientific knowledge of something then I greatly prefer it to any kind of, let's say, lesser knowledge, certainly to anything like hearsay. But when I talk about my own experience, that is something of a similar value as scientific knowledge. I didn't need to prove to myself that lucid dreaming was real; you didn't need to prove it because you had the experience; so the third person scientific proof was only necessary for people who didn't have it.

Eastern traditions have been working at this inner knowledge for thousands of years. And I think that we in the West have the unique opportunity of benefiting from an interaction with that Eastern tradition, bringing in the Western scientific perspective. I think the collaboration of these perspectives is what will give us the potential to understand consciousness in a new way, and then to make use of the value that it has, in becoming fully what we might be.

Sue At the moment there seem to be two groups amongst people studying consciousness. There are those who are doing it very much from an objective point of view, studying the neural correlates of consciousness or doing brain-scan studies, who are not, on the whole, interested in self-transformation. And there's another bunch of people who are interested in altered states and Eastern religions and so on, and who are somewhat antagonistic towards the hard science. What do you think is going to happen?

Stephen I think we need a third possibility. We need to have scientists who understand the brain but also have their own experiences. The problem to be explained is experience, and if we believe that the brain is the way we're going to understand experience, but then try to study the brain without studying experience, then what are we explaining? To me, the two naturally go together. I'm interested in both ways of looking and understanding. That's why I'm a psycho-physiologist; it's exactly those two corresponding perspectives, the inner view and the outer view, that really fulfil my life. I wouldn't want to give up either of those two approaches.

Sue Do you get much antagonism from other scientists?

Stephen Of course, if you take this point of view, you're going to get it from both sides. If I'm talking to New Agers then I've got this weird scientific attitude that they just don't understand. And if I try to talk to scientists who don't have any sense of the experience, then they think, 'This guy must be weird, what's wrong with him?' But there are other people who will understand or will have the same experience for themselves—as I'll tell the world that Sue Blackmore has had some very curious experiences and that it's made her a very interesting person and opened her mind to novel ways of looking at the world.

Sue It seems we share the hope that these two will come together.

Stephen Yes, and I think we're going to see it happen.

Sue Do you believe you have free will?

Stephen That depends on what we mean by free will; and it depends on what we mean by will; and it depends what we mean by me. If you mean, 'Does my conscious mind, that model of myself, the one I was talking about in the dream, decide what it wants to do, or how it's going to answer this question?', then, no, I don't think so. But 'Does who I am, and all that I am, decide how to answer this question?', then yes. The problem here is, what do you mean by me? When I've got free will, what's the 'I' that's got it?

Sue When you were talking about enlightenment, you described it as something like the individual or self, slipping into a great unity. Could you say, in the question of free will, that the choices are coming not from this little conscious you, nor even from this body, but from everything?

Stephen Yes, and that's why it depends on what you mean by 'me'. When I speak of the totality that I am I don't mean just this complex body stuff here; what compelling reason do I have to limit it to that? Given the experiences I've had, I have to keep an open mind on the question 'What am I?'

Thomas Metzinger

I am the content of a transparent self model

Thomas (b. 1958) studied at Johann Wolfgang Goethe-Universität in Frankfurt, did his doctorate on mind-body problems at Frankfurt University, and since then has taught at several universities in Germany and the USA. His philosophical interests include ethics, the nature of self, and the philosophy of science and especially of cognitive science and neuroscience. He is best known for his self-model theory of subjective experience, and is a long-term meditator. He is Professor of Philosophy, and Director of the Theoretical Philosophy Group at the Johannes Gutenberg-Universität in Mainz. He has edited *Conscious Experience* (1996) and *Neural Correlates of Consciousness* (2000) and is author of *Being No One* (2003).

Sue What is it about consciousness that makes it so special, so different that people talk about 'the problem of consciousness'; what is the problem?

Thomas The problem is that consciousness is opposed to all other states. A physical state, a biological state, a chemical state are only known from the outside, from the third person perspective. Consciousness is different in that we gain knowledge about it from the inside as well as from the outside—and we don't really understand what that statement actually means. Consciousness, we say, is also known from the first person perspective, by an experiencing self. Therefore, there are two different ways of gaining knowledge about the phenomenon. Philosophers call this the 'epistemic asymmetry'—philosophers always have difficult words for things like that. You can get objective third person knowledge by looking at the physical properties of a human person, for instance—by looking at that person's body; and you can in principle get complete chemical knowledge about all the chemical states that make up the body of that human being; you can describe the biology of human beings, and you can also describe the neurobiology, that is, all the objective scientific properties; you can describe the brain of this person—you can even move up to higher levels of description: you can describe how their brain currently processes information, what kind of representational contents (another philosophical term) it activates. But when you come to the interesting level, the level of conscious experience, and you want to gain knowledge about the conscious states of that person, you suddenly find that there are two ways of accessing these states: one is from the inside, from the first person perspective, which can be used by that single individual person herself; and the other is by accessing the physical correlates; whatever happens in that person's brain.

So it is the only natural phenomenon—and I'm convinced that it is an entirely natural phenomenon—that can be known from the inside and from the outside, and the problem is that we don't have a proper understanding of how that inside and that outside are related, and in particular what we're actually talking about when we say something like 'knowing from the inside'.

Sue I've been talking to some people such as Francisco Varela and Max Velmans, who say it isn't really like that at all, that there isn't really a difference between looking at things from the inside and looking from the outside; that if you take a wider perspective it's all the same thing. Presumably you disagree with them?

Thomas It depends on the level of description. What we most urgently need to know is, what is a first person perspective? We need to know what it is that makes my own conscious state what it is, how I appropriate it as a conscious self. This thing—subjectivity—we loosely talk about it all the time, but we don't know what it means. I think I know what it means.

Sue Do you? I think I know what it means in the sense that I can look out of this window now at this beautiful copper beech tree, and feel that there's a private experience going on here, from this viewpoint that nobody else can ever know about, and that it's very rich and vivid and colourful for me, and that you must have another such experience that I can never get at. And that seems to me to be a mystery. I don't know what to do with that mystery; do you?

Thomas It is certainly true that conscious experiences take place in individual models of reality, in individual brains, and from an individual first person perspective. The question is whether you can generate a real, deep mystery out of that fact. And the deeper question is, what the hell is a self, and what actually is a first person perspective? I would argue it is a very, very specific kind of representational structure, a way in which brains depict the world as a centred world, as a world that's centred around a self, and which has proved to be adaptive, biologically successful. That's what generates the problem.

Related to that, I don't think there is a principled explanatory gap between the brain and conscious experience, but there may indeed be something like an intelligibility gap. It may be that even if we have a satisfactory theory of consciousness, this theory is not intuitively plausible to us, and we cannot consciously experience the truth of that theory. The funny thing today is that if a physicist comes along and tells you something about eleven-dimensional models of reality, and string theory, and how the universe started before time started and fancy stuff like that, nobody says, 'Well, this is intuitively implausible, I cannot perceive this; this is not a good theory because it's not intelligible to me.' On the other hand, everybody thinks a theory of consciousness has to immediately give you some intuitive insight into the phenomenon. But nobody ever said that a theory about a phenomenon should recreate the phenomenon—so for instance, a theory about bat consciousness and the sensory qualities of bats does not have to create that consciousness for people who read that theory in books later on. This is just a false criterion for what would be a good theory about consciousness.

Sue You're implying that because we all think we know what consciousness is, we don't like theories that don't immediately gel with the way we experience the world. Do you think, then, that perhaps we need better training; that we as scientists investigating consciousness need to train ourselves better to look at our own consciousness and to be aware of its capacities and its differences, and that that might help us to a broader understanding of all these theories?

Thomas Obviously it would be at least heuristically very fruitful if scientists working on the problem could have rich and enhanced first person access through a wide variety of alternate models of reality—if they knew what was happening in meditation, or during trance dancing, or drumming—but that wouldn't necessarily give them good intuitions or better ideas. Progress would be faster, I think that is certain, but that doesn't mean that the final theory we arrive at would be intuitively more plausible even for those scientists.

For instance, to take a rather simple point, such a theory might say that there are no such things as colours in the world, colours are not objective properties of things out there; and you can start believing that, even though you will still *experience* them as external. But that theory might also say something like, 'No things as selves exist in the world; what we call our experience of being someone is in itself a complicated representational phenomenon'—actually our best theories might in the end say something like that. How should we conceive of this as we are conscious systems operating under such a phenomenal self, under such a phenomenally experienced first person perspective, which somehow we cannot transcend on the level of experience?

Sue But that's a really good example—the idea of self—because it seems to me, intellectually, that there probably isn't a self in the sense of a persisting entity that is the subject of all the experiences. And that is intuitively very difficult to handle. Nevertheless, undertaking meditation or mindfulness practice, makes it much easier to cope with the idea that there isn't any self, and so it seems to me it would be useful if we would train ourselves more to accept that.

Thomas I think it would be useful if we would train ourselves—and our children—to that, and I have been doing that myself for almost three decades now, but of course it cannot be an obligation for people to do so. I think a simplified but fruitful way of looking at things is to say that we are never in contact with reality as such, and we know reality only under representations.

There are two kinds of representations. There are theoretical representations, like knowing about consciousness within a certain theory that brain scientists or psychologists have made. That is one way of gaining knowledge about consciousness and what you really are. It is stored in books, computers, and ongoing scientific discourse. Another way of accessing reality is through a phenomenal representation, in the way your conscious mind, your brain, happens to depict reality and yourself. Scientific representations of the world, and of consciousness, aim at maximal objectivity, at being very parsimonious, at not introducing superfluous entities, and at making good predictions. Phenomenal representations are clever in a different way because they had a completely different purpose: they were needed to help our parents and grandparents and all our ancestors to survive and copy their genes. Their target was not to generate a faithful representation of reality or of the brain, or the way we sensorily perceive the world; they had a completely different goal, and certain illusions can be functionally adequate—as philosophers say of misrepresentations: the belief in your own existence as a distinct self or, to say something more provocative, the belief that life is actually worth living, can be very successful in copying genes.

Sue You seem to be saying here something quite weird: firstly that there isn't really such a thing as the self in the way many people think there is, and secondly that it has been biologically useful for the genes to construct this illusion. Can you explain that?

Thomas A maximally unromantic and sobering way to look at the content of self-consciousness is to look at it as the content of a transparent self-model, as philosophers would say. There is an internal image of yourself that you cannot recognize *as* an image while it is there—and the unromantic part is in regarding this as a weapon that emerged in the cause of the cognitive arms race. There was constant competition among organisms on this planet, for millions of years, and it was merciless and cruel and the development of things like memory, thought, better perceptions, was just as important as better legs, better livers, better hearts. I like to look at the human self-model as a neurocomputational weapon, a certain data structure that the brain can activate from time to time, such as when you have to wake up in the morning and integrate your sensory perceptions with your motor behaviour. The ego machine just turns on its phenomenal self, and that is the moment when *you* come to.

To have a good self model means to be successful in a certain environment. It starts with very simple properties: you need to know how far you can jump, what your body can do, how big you are, what your boundaries are, so that you don't start to eat your own legs, as some primitive animals may actually do, or as some psychiatrically disturbed people do. The question is what makes a self model a good self model? It can be appropriate in having a lot of children and grandchildren, it can also be appropriate relative to a certain social environment.

If you're drunk with your friends on Saturday night at a quarter to three, you usually have a different self model from when you're visiting your parents for breakfast at 10 o'clock the next morning. So it can actually be a sign of mental health to have variance in your self model, to have different self experiences, different phenomenal conscious identities in different social contexts; but it can also get out of hand, as you can see in multiple personality disorders—or in politicians.

Sue So evolution has played yet another trick on us. It has not only given us bodies that are determined to stay alive even if we find it rather painful being alive, but has caused us to produce a false or misleading sense of self, which we'll go on defending because it's useful and because it helps pass on the genes of our ancestors.

Thomas Well, first of all evolution is not an agent itself: evolution doesn't play tricks, it is simply mindless, merciless self-organization. It just happens like this on this planet in the universe. Another thing is that I cannot believe—if it is true that phenomenal states, conscious experiences are representational states, images of reality—that all of it is false. There is an external physical reality and there is an internal reality of the body: your temperature, your blood sugar, your emotional state. So in most cases there must be some faithful depiction, otherwise it couldn't be successful in managing the physical body and in navigating a physical environment. But it may be that there are certain higher-order features which are particularly illusory from a strictly objective or philosophical point of view—for instance, the experience that you are an enduring entity, or that there is some essence of you which is invariable across time. But then again, if you leave scientists in their academic circles and talk to good normal people who really have common sense, they all know that you are not the same person across a whole lifetime—actually we have all known this for centuries. It is also a question of how we actually describe ourselves. If we are Christians, or if we are Cartesian philosophers, then we have a certain way of describing our conscious experience; we look for an enduring self, and then we find it.

I think one task may be to go, with introspective attention, into the real, deep structure of conscious experience without making theories, without naming things, without relating them to anything in the past, and to see whether there is anything like selfhood as such there, independent of all descriptions or whatever beliefs or pet ideologies we may happen to have.

Sue And have you tried that? And what have you found?

Thomas Well, the big problem in the process is the person who tries. Wittgenstein has already remarked that all those people to whom the meaning of life became clear have not been able to say what it consisted in, and in many cultures you have these old sayings, proverbs, like 'Those who know don't speak and those who speak don't know,' that kind of thing.

That also makes it difficult if somebody comes along and says, 'Yesterday I woke up early and I walked to the forest and I sat down and suddenly—you won't believe it—I became one with the world, and I felt—my God—actually I was not there at all, and it was a selfless universe.' From a methodological point of view such reports are very dubious, because if you were not there how did the autobiographical memory get formed, how could that have been an episode of your own life that you can now report? So it's very hard to make sense of these reports if you have a more rigorous perspective on it: do people just report that something happened and then put a theory on it which they have had before, or heard or read—what is the true fact of the matter?

But then maybe there are areas in human life where the point is not, as in science or philosophy, to find out the true fact of the matter; maybe there are areas of life where you should just rest in effortless attention and dissolve in the present moment, and there is no reward to be gotten, no message to be brought home; this could be true too.

Sue Do you believe in the possibility of the philosopher's zombie?

Thomas I am not a possible-world surfer. As long as 'consciousness' is such an ill-defined term, many things remain conceivable. The zombie-thing is an expression of the time we now live in—200 years from now zombies will not be conceivable any more. Today, I don't think that we can make substantive progress and advance our knowledge and understanding of consciousness by kicking the problem upstairs into formal semantics and modal logic. But I may be wrong.

Sue What about free will? Do you have free will?

Thomas If I didn't, could I ever have given you any other answer than this one?

Sue You've spent years and years thinking about the philosophies of consciousness and about models of self and so on. What has it done to your ordinary everyday life? Is it separate, do you go to work and do this

thinking and go home and forget it, or is this really deeply mixed up with your own life?

Thomas It is deeply mixed up with my own life, and in particular I think I'm paying a price for doing this kind of research. For example, I often study neuropsychological syndromes, people who have severe brain lesions, or people who have gone mad, and I analyse these states as a philosopher—but of course you also always try to understand how it really feels like to be such a patient, and if you really do that it hurts, and it makes you become aware of the fact that any time—when you're walking across the street—some little thing might happen in your brain to completely deprive you of your dignity for the rest of your life, and turn you into one big suffering confused mess. We are very fragile beings.

I also don't think that in general an academic career or an academic life is something that makes you particularly happy or is conducive, say, to meditative states. You have a lot of hypocrisy and competition, hard egos, and particularly clever and ambitious examples of the human species. It's not such a beautiful, social environment to live your life in—but it is, of course, very exciting to follow the old philosophical ideal of self-knowledge and to be ready and have the guts to really face the facts, and to make use of the enormous new tools we have in cognitive neuroscience right now. But what I think many people, including many professional philosophers, don't understand is that nobody ever said self-knowledge is emotionally attractive, or that it cannot also have sobering or outright depressing effects on you.

There are hard theoretical issues, which you can only talk about with philosophers and scientists, but there are also what I call 'soft issues' and these soft issues have been making me more and more concerned recently, because I think something is coming towards us as mankind, and it's coming very fast, and we are not prepared for it.

Let me give some examples. There is a new image of man emerging out of genetics and neuroscience, one which will basically contradict all other images of man that we have had in the Western tradition. It is strictly unmetaphysical; it is absolutely incompatible with the Christian image of man; and it may force us to confront our mortality in a much more direct way than we have ever before in our history. It may close the door on certain hopes people have had, not only scientists and philosophers but all of us, such as that maybe somehow consciousness could exist without the brain after death. People will still want to believe something like that. But just as people will actually still think that the sun revolves around the earth—people whom you basically laugh at and don't take seriously any more. So there's a reductive anthropology that may come to us, and it may come faster than we are prepared for it; it may come as an emotionally sobering experience to many people particularly in developing countries, who make up 80% of human beings, and still have a metaphysical image of man, haven't ever heard anything about neuroscience, don't want to hear anything about neural correlates of consciousness, want to keep on living in their metaphysical world-view as they have for centuries.

Now here we come in these rich, decadent, non-believer Western countries, and we suddenly have theories which work very well in medicine and in treating psychiatric disorders, and which say 'There is no such thing as a soul,' and 'You are basically a gene-copying device,' and it is not clear what that will do to us. A chasm will open between the rich, educated, and secularized parts of mankind on the planet and those who for whatever reason have chosen to live their lives outside the scientific view of the world, and outside the scientific image of man.

Our image of ourselves is changing very fast, but there's a problem associated with it: that image, in a very subtle way, influences the way we all treat each other in everyday life. One question is, for instance, whether a demystification of the human mind can take place without a desolidarization in society. What has held our societies together and has helped us to behave have been metaphysical beliefs in God or psychoanalysis and other substitute religions like that.

The question is, can science offer anything like that to keep mass societies coherent after all these metaphysical ideas have vanished, not only in professional philosophers and scientists, but in ordinary people as well? If everybody stopped believing in a soul, what effect would that actually have in the way we treat each other? All this may have cultural consequences which are very hard to assess presently; it may have a broad effect on the way we view each other, and it is very important that a crude, vulgar kind of materialism is not what actually follows on the heels of this neuroscientific revolution. For this, transported through the media, makes people believe in simplistic ideas such as that human beings are just machines, and that the concept of dignity is empty, and there never has been such a thing as reason, or responsibility.

Then there is another class of soft issues: what do we do with our new image of ourselves; are we ready to face the facts? Another set is what I call 'consciousness culture': what is the impact of all this on ordinary everyday life for all of us? And the third is what I call 'consciousness ethics': as soon as we know more about the brain and the neural correlates of consciousness, we will at least in principle be able to selectively switch conscious experiences on and off with new molecules, or by using direct transcranial magnetic stimulation, to create new media environments in the global data-cloud, new forms of electronic entertainment that we have never dreamt of—cyberspace worlds, holographic cinema, etc. And then there is all this info-smog and increasing speed in the business world, which is already too much for many of us today.

We all realize, now that the Internet is humming all around us, that in one way it's a blessing and it helps us, and that in another way it enslaves us. To give you an example, I recently became aware that I was in a dream, and I realized that by the fact that the transition from one dream scene to the other looked exactly like the way I click from one website to another. So, working with all these computers and new technologies does something to the brain itself.

And another thing, drugs: we're going to have terrific biological psychiatry, terrific medicines, in 50–100 years' time, to get rid of things that have plagued mankind for millennia. On the other hand, we will also probably have recreational drugs that mankind has never dreamt of. So if, for instance, we could have something that is non-addictive and has no major side-effects and puts a nice smile and a sexy flirt on to our faces, and you can take it for three decades. And if your doctor says, 'What you have is only a common sub-clinical deep depression; you're not getting this,' people will say 'I am a free citizen. This is my brain. Why does the medical profession have the right to tell me how I am going to design my conscious life?'

I want to be an autonomous person in that open future society. Making these things illegal will not help, because wherever there is a market there will be an illegal industry which serves that market. So the times where we were wondering about the neurotoxicity of Ecstasy and things like that may actually look like an Easter Sunday walk to us in ten or 100 years when children and adolescents are coming to psychiatric emergency wards under the influence of substances the doctors never heard of when they studied medicine at the university, because everything is flooded with ever newer molecules and more and more efficient ways of changing consciousness. The old strategy—laws, disinformation, and repression—will not do in such a situation: either we find a sane way to use all these new tools in a mature and intelligent way or we will be in big trouble.

Sue Are you optimistic about the future yourself?

Thomas Personally, I tend to be rather pessimistic about the future, but I always remind myself that of all those super-well-paid sociologists and political scientists in all the universities around the world, not one of them predicted that the Berlin Wall was going to come down, and everyone of them could have made a big career and become very famous by predicting that. So I think that human history is basically open and none of us really knows what's going to happen next.

Kevin O'Regan

*There's nothing there
until you actually
wonder what's there*

Kevin O'Regan (b. 1948) studied Mathematical Physics at Sussex University and then at Cambridge where, after two years, he switched to Psychology and a PhD on eye movements in reading. He has studied word recognition, change blindness, and the stability of the visual world despite eye movements and is trying to understand the phenomenal experience associated with sensory stimulation, in part through developing a sensorimotor theory of vision. Apart from science, his favourite activity is Capoeira. He is Director of the Laboratory of Experimental Psychology, Centre National de la Recherche Scientifique in Paris.

Sue **What do you think the problem of consciousness is?**

Kevin   A pseudo-problem.

Sue   Aha.

Kevin   No problem at all.

Sue   Go on—when people say it's a pseudo-problem or it's no problem at all, it seems to me that either they just don't get it, or they've really seen through the problem and understood something. I suspect you're one of the latter, so can you help me to see through it?

Kevin We all have the really intimate experience of living, and seeing things, and thinking, and believing we exist and so on, which needs some kind of explanation; and as scientists we all believe that this experience comes from some brain process; the problem is making the link between the experience and the brain process. And nobody seems to have found any reasonable physico-chemical mechanism that could make that link.

Sue So isn't that a real problem? You've articulated there the hard problem; why do you say it's a pseudo-problem?

Kevin Because it comes from a misapprehension of what experience really is. At the beginning of the century people thought there needed to be some vital essence that endowed living organisms with life, because they thought life was a unitary magical thing. Then a paradigm shift occurred; people realized gradually that each aspect of life's components can be accounted for by some simple mechanism which had a physico-chemical explanation—a materialist, no magic explanation. And gradually people simply abandoned the idea that life was a unitary thing with a 'vital essence' that generated it. I think exactly the same paradigm shift could solve the problem that people call 'the hard problem of consciousness'. If they simply abandon the idea that consciousness is a single thing that is generated, or that emerges from some brain process, and simply look at all the things experience really consists in, then each of those things will have its own little explanation and there will be no need to invoke any new magical process.

Sue But it doesn't seem like that, does it? It seems to me—most of the time, anyway—that here is this unitary world, and here I am, experiencing this room in all its richness. How can that be similar to the problem of life?

Kevin I think that *you* would say that it's a meme, that's to say the idea we all have that we really are experiencing stuff and that it really exists; that we ourselves are acting, existing beings with free will and with raw feel. These are things we say to each other and that we convince each other of, but we could just as well change our view on this and say something different.

I remember as a child reading books written by physicists at the beginning of the last century, about the origin of life, saying what a wonderful, mysterious thing life is, and asking whether perhaps the origin of life derives from something special in proteins, and I don't know what. If you read those books, it seems that people had what you would no doubt call a meme about the nature of life as a magical thing that emerges in living organisms.

Sue **You're right, that is what I think about the self; that it's a story built up by a collection of memes all getting together for their own benefit, not for ours. Nevertheless, although I can say that intellectually, inside it still seems to me that I am having a unitary experience. Yes, in meditation it begins to fall apart, and when I look introspectively with a calm mind I don't see a self looking; nevertheless it seems to me in ordinary life that there's a continuous perception of the world. Are you saying there's something wrong with that, that it isn't like that?**

Kevin That's right. Although we all have the idea that we have a continuous perception of the world, my claim is that that's an illusion. I can give you a couple of analogies that might help you understand the origin of this illusion. One example is what we call the 'refrigerator light illusion', which was suggested by the philosopher Nigel Thomas. The idea is that when you open the refrigerator, the light's on; then you close the refrigerator and you wonder, 'Well, is it still on?', you check by quickly opening the door and it's on again. So you have the impression that the light is always on. Similarly with the visual field: you have the impression that you see a wonderful rich field in front of you, and that the world is continually present. I claim that actually it's not; there's nothing there until you actually wonder what's there, and it's the fact of wondering that allows you to perceive it—rather, I should say, that *consists in perceiving it*.

Sue **Since you first put this analogy into my head a few days ago, I realize I've been going around trying to catch the world out. I quickly look round and of course you can't catch it out, however quickly you open and shut the fridge. It does do funny things to one's perception. But you said you had other analogies.**

Kevin Well, part of what people find hard to take about my view is the notion that seeing is not something ongoing and occurrent in the brain. Most neuroscientists are looking for brain activation as a correlate of experience; they're looking for some brain area that lights up and would be the cause of visual experience. But my claim is that nerve-firing couldn't possibly be the experience of seeing, because if it were you'd have to postulate some magical physico-chemical mechanism that took these neurons and their activity and translated that into something which is essentially non-physical, namely experience.

So you can get out of the problem by the simple trick of saying that experience is not *in* the brain; it is something that the brain can *do*; it's a brain capacity. Now of course there is an underlying nerve process going on to provide this capacity, but the capacity doesn't consist of nerves firing, it's something the neurons allow the organism to do when it's put in the appropriate situation.

So another analogy is with being rich. If you're rich you know that if you go to the bank you can take out a big wad of money; you know you can write a cheque and go off on an expensive cruise—so the feeling of being rich is not something going on in your brain, but rather a capacity: it's knowledge that you could do something if you wanted to. On the other hand *at any particular moment* you can have the feeling of being rich without doing anything at all.

Sue **In what sense is vision like that? Are you saying that what's happening in vision is simply that I have the illusion that all this world is here, but only because if I want to I can look again?**

Kevin It's not that if you want to you can look again, because it's not the fact of looking at something that gives you the experience; it's the fact that you know that if you should move it would provoke a certain change in your sensory input.

You mustn't think that what gives you an experience is the fact that you move your eyes and look at something new; you shouldn't think of it as a snapshot, coming in when you move your eyes. It's knowledge of the change that occurs when you move your eyes, or move your body, or move that object; it's knowledge of the changes that would occur if you did something—but remember, you don't actually have to do anything.

Now, you might say there's a bit of a problem here because richness doesn't really have a feel at all. That's rather interesting because in fact vision is quite different from being rich in that it's much more intimately linked with your actions. If you blink, the visual field goes blank; the slightest twitch of your eye muscles or your body provokes an immediate change in the sensory input that you get from your eyes. But if you blink nothing happens to your bank account. So there's an incredible difference between the feeling of being rich and the feeling of seeing, and I think that the advantage of my theory is that it gives a principled account of such differences in feelings. You have the feelings of being rich, of being happy, of being poor—these are more conceptual types of feeling. Why? Because they're not intimately linked to action. Whereas you have perceptual feelings like seeing, hearing, smelling, which are intimately linked to certain types of bodily motion.

Furthermore, this theory explains certain mysteries that over the last century nobody has satisfactorily explained, in particular why it is that you hear sounds and see sights and smell smells, instead of, say, *seeing* sounds, *hearing* smells, or *smelling* sights. The classical view is that you see sights because they stimulate the visual pathways—but that doesn't explain anything. The visual cortex has a quite similar structure to the auditory cortex; there's no reason why neurons in a visual cortex should give rise to the particular feeling associated with vision. Appealing to brain processes to explain the differences in the different sensations puts you in the terrible situation of having to postulate some magical mechanism that endows visual cortex with sight and auditory cortex with hearing.

On the other hand my view provides a perfectly simple explanation: you have the *seeing* feeling when certain types of changes are associated with certain types of movements. For example, you close your eyes, the visual field goes blank; you move forward, it creates an expanding optical flow on your retina. However, closing your eyes has no effect on the auditory information. You know you're hearing if you move your head and the asynchrony between the incoming information in your ears changes in a lawful way. Hearing and seeing are knowing that certain laws apply.

Sue **Do you think this is how we distinguish real vision from imagination?**

Kevin Yes, and hallucinations and dreaming perhaps, because when you blink nothing happens, or at least you don't get the right laws being obeyed as with normal seeing, although there is nonetheless a lot of similarity.

This also explains phenomena like sensory substitution. There have been studies showing that if you connect an array of vibrators to a television camera through some electronics, such that the pattern of vibrations forms a tactile image of what you're looking it, then if the blind person who uses this apparatus moves the camera around he can actually come to have a sensation similar to seeing. He can have the impression that objects are outside him.

Sue **Does this give one some insight into what it's like to be a bat?**

Kevin It certainly says that we have little hope of really knowing what it's like to be a bat, because vision is intimately related to the particular laws that our visual apparatus possesses. So, for example, when you, as a human, look at a red bit of paper directly, it's being sampled by your foveal vision, which has a high density of colour sensitive photo-receptors. But when you look slightly away, the density of colour-sensitive photo-receptors is much less, and so the quality of the incoming red light changes. What's more, there's a kind of bluish greenish goo on the central part of your retina that filters the red light in a different way from what happens when you look away from it. So actually there is no such thing as red. Redness, in my view, is the way red surfaces change the light when you move around with respect to them.

Sue It's very interesting to think about red. Red is often used as the paradigm example of a quale; the raw experience of the redness of red. Yet we know from psychology and neuro-physiology that objects aren't of themselves red, that red emerges in an interaction between light and an eye and a nervous system. What you said there makes it even more complicated, because red is one kind of thing when you're looking straight at it, and it's different when you move your eyes around. But I find it very difficult to cope with this in my own experience. I mean, here I am looking at the world, which seems solid and full of red and brown and green objects. Are you saying I'm wrong?

Kevin I'm saying that our experience is not due to a brain process; it's constituted by a brain capacity. Looking at it that way explains the nature of experience. It explains why being rich has less of an 'intimate' feel than seeing a red patch; it explains why seeing is different from hearing, and it explains things about pain. When you have a pain your attention is incontrovertibly drawn to the painful stimulus. Not only is your attention drawn to it, but you can't remove your attention from it, and so that gives an additional difference in the raw feel of pain as compared to other perceptual experiences.

Sue So in your view, understanding what experiences are like must necessarily involve understanding what you can do with them?

Kevin Yes, almost—except I would disagree with you in the sense that I would say experiences are what you can do; you mustn't say 'understanding what you can do with them,' because ... because experience is what you do, it's something you do.

Sue Are there any ways in which you can test this theory?

Kevin I've been doing a series of experiments on what we call change blindness, where I show that even though you have the impression of looking at a very rich visual scene, under some circumstances I can make enormous changes in those scenes without you really noticing them. This casts doubt on the idea that most neuroscientists and psychologists have, that we have an internal representation of the outside world. So that's one consequence of this way of thinking: that we don't re-present the world inside our brain. On the contrary, we use the outside world as a kind of outside memory to probe. There's no need to make an internal replica of the outside world.

Sue I seem to be nearly getting what you mean, and often getting it wrong, so I want to be really clear about this. Is what you're saying something like this: the way we tend to think in neuroscience is that we open our eyes, look at the world, make a rich internal representation, and that's our experience. And you're saying something like, 'No, it isn't like that, we haven't got this rich representation, actually the information is stored out there in the world, and we just get little bits here and there in a sort of fragmented way,' something like that?

Kevin Almost, almost.

Sue I'm only almost again!

Kevin Almost because you use the word 'get', you said 'we get little bits of information,' and that suggests that we're getting these bits, and we're putting them into our brain, which is wrong. The actual experiences that we have derive from the activity that we have within our environments.

Sue I've really got to make some fairly dramatic changes in my way of thinking about vision, haven't I, in order to understand what you're saying?

Kevin I'm surprised at you, because you are one of the people with the closest views to mine.

Sue Well, I think I'd better get them a bit closer—and then I'll have to decide whether to stay that way or reject them. Now go on, explain the basic idea of a change blindness experiment and how it works.

Kevin Well, I'll tell you what the experiment looks like first, and why people are surprised; then I'll give you the explanation. In the experiment I show a picture to somebody and something enormous changes in the picture. For example, let's say it's a picture of a Paris street scene with Notre Dame occupying say a third of the background of the picture; and then suddenly I shift Notre Dame about a quarter of the way across the picture. Now normally you'd see this immediately. But what I do is, just at the same time as I make the shift of Notre Dame across the picture I put a very brief blank in between—it lasts maybe two or three tenths of a second—so it looks just like a sort of blink. If you do that, people just don't see the changes; and then when you say to them 'Ah, can't you see that Notre Dame is moving,' they say, 'How could I have not seen it?' It's totally obvious when you know what you're looking for, so people are very surprised.

Sue **Let me get this straight: you show a picture and there's an enormous change. Normally people will easily see the change; but if you insert a flash or a gap, or move the picture as I did in my own change blindness experiments, they don't see the change, even though it's huge.**

Kevin … and even though they might actually be looking directly at the location of the change.

Sue **Right, so what do you think is going on here?**

Kevin What's going on, in my explanation, is that since we don't have any internal representation of the picture, we're using the picture itself as an outside memory. The only way we can know that something has changed is by having remembered that bit, and we have very poor memory. You know, if you look at a picture and close your eyes, and ask yourself what exactly was there, you'll probably be able to say, 'Well, there was a table, there was a chair, there was a bed,' but if I ask exactly what part of the table the bottle was occluding, or what exactly was the pattern on the bedspread, you won't be able to tell

Change blindness. When these two pictures are swapped just at the moment when someone blinks or moves their eyes, then the person does not see the change. The same effect can be obtained by alternating the pictures with a brief grey flash in between. This suggests that we do not store a rich and detailed representation of the visual world that lasts from one eye movement to the next.

me. You have a semantic description of the picture, which is similar to what you would get if you read a description in a book.

So my claim is that one's internal knowledge about what one is looking at is essentially nil; it's limited to a simple semantic description. But when you have your eyes open this description is enriched by the visual stuff out in front of you, and you have the feeling that this is real. And so if there should be any bit of the visual scene about which you want to have more information—for example, the exact pattern on the bedspread—all you need to do is look at it, and the slightest flick of attention, or of your eye, will immediately make it available.

It's like the refrigerator light analogy again: you think that you're seeing all the colours on the bedspread because if you should merely wonder what they are, your eye will go there and you will know.

Sue So what you're saying is that while you're looking at something, you have rich information available, but that every time you move your eyes, which is four or five times a second, the information's just gone. There's no solid representation, there's no visual memory that persists between the...

Kevin You said something wrong there.

Sue Oh, I'm always doing that, but I really want to understand this, so correct me; what have I said wrong?

Kevin You said 'it's available'. That's wrong. It's not available. You mustn't think that when you're looking at the bedspread the colours of the bedspread are impinging upon your retina, activating some internal representation of the colours of the bedspread. That's not what's happening. What's happening is that if you're looking at the bedspread, and you wonder, 'Is that bedspread chequered or has it got a tartan?', then you will do the appropriate investigation of the bedspread to answer that question. I'm saying that you don't see anything other than what you are interrogating yourself about. There is no real seeing; there is no seeing other than that which corresponds to your interrogation of some aspect of your visual environment.

Sue So with every saccade, every natural eye-movement, I'm interrogating the environment in some way, I'm attending to something or another—that's the experience. But then as I move the eyes again, or think about something else in the environment, all that's just gone and a new thing's started up.

Kevin Right. What remains after each eye movement is a semantic description of the scene—something which is essentially non-visual.

Sue This is a very weird way of thinking about the world, but it begins to make sense. And those experiments suggest that it's right; that what I've got is a very sketchy semantic description, or conceptual idea: 'Here I am in this room and there's a bottle there, and what have you'—backed up all the time by the process of interrogating the world, looking to this, paying attention to that, and so on, all of which is not saved: it doesn't hang about as a representation in my head; it's just there in the world being used or not being used.

But that's not how it feels to be alive and seeing, is it?

Kevin If you think about it, it does. If you consider that seeing is the fact of testing the laws of sensorimotor contingency between the sensory input and the motor output—if you just change your vocabulary, change the way you think about what the experience of seeing is, forget about your meme that it consists of creating an internal representation, and see seeing as an activity—then it makes sense.

Sue What does this view say about the me who's doing the looking?

Kevin Well, I'm not a philosopher. I have my own view on the question which probably would be labelled behaviourist, or neo-behaviourist, or perhaps somewhat Dennett-like, or even compatible with your ideas on memes, which is that the notion of self, the I, is merely a social construct which allows me to conveniently describe the things that 'I', in inverted commas, do.

Sue So you're implicitly distinguishing between I, this whole body or this whole system that is sitting here now, and the 'I' in inverted commas which has this sense of being aware, being conscious, being in control?

Kevin I'd say that that sense is only a way of talking about what I'm doing; it is again a kind of social illusion, a practical way of thinking about things and of talking about things to others. But there's no magic about it, there's no extra mechanism that has to be added into a human brain in order to get him or her to feel this selfness or this consciousness.

Sue You've used the word illusion a couple of times. I think that people get into a lot of confusion about this, either they think that illusion means that something doesn't exist at all, or they have some other odd idea. What do you mean by an illusion? And how much of our experiences and theories about ourselves do you think are illusory?

Kevin   I agree that the term illusion is a bit tricky. I've been chastised by philosophers who claim that my use of the word illusion is, well, illusory! They say that it makes no logical sense to say that vision is illusory, because vision is vision, and vision can't by definition be illusory. When I use the term, what I mean is that vision is not what you think it is, or what today's psychologists and neuroscientists and philosophers think it is. I essentially use the word to shock people.

Sue   But that may be right. If you're saying that most of us think of vision as being an experience of a world that's constructed inside our heads, and that we're wrong about that; then it is an illusion—vision isn't what we thought it was.

This reminds me of something that happens so often in the field of consciousness studies, and probably elsewhere too: that you take something like our ordinary everyday assumptions about the world, and challenge them, and then everybody says, 'Oh, I knew it wasn't like that really.'

Kevin   It's rather amusing that when I sent my paper on mud splashes to *Nature*, there were two reviewers: one said 'This is brilliant,' another said, 'This is banal,' on the grounds that it was a well-known fact that one had no visual memory. The editor of *Nature* said, 'We cannot publish a paper where the reviewers are divided, so we reject.' I wrote back with what I thought was an excellent gambit, which was to say, 'Look, your two reviewers disagreed, and indeed the whole scientific community is divided: half of the people think it's brilliant and half think it's banal. There is an important scientific controversy here; how can *Nature* stand apart from this scientific controversy?' And it got accepted.

Sue   This was where instead of a gap in between the two pictures you put little splashes, as you might get when driving a car with mud splashes on the windscreen. The implications of this for actually driving cars are quite worrying, aren't they? It suggests that if you get a big splash on the windscreen you might not notice an important change like a child running out in front of you, or a bus.

Kevin   Absolutely. It doesn't even have to be a big splash, just a few little splashes would do.

Sue   You've been talking here about the experiments you've done, and the dramatic changes in the way you think about vision. How much of this has changed your life, your way of relating to people, your way of going about ordinary living?

Kevin It hasn't changed it at all, because I knew I was a robot, and I was just trying to prove it to people. And finally I've managed to get it across to them.

Sue Tell me about that; how did you come to know you were a robot? Did this start from birth? Were you born a zombie?

Kevin Ever since I've been a child I've wanted to be a robot. I think one of the great difficulties of human life is that one's life is inhabited by uncontrollable desires and that if one could only be master of those and become more like a robot one would be much better off.

Sue Ah, this robot is a different sort of robot from the one I imagined: this robot has control over its desires, does it? I imagined the robot might be more like Data from *Star Trek*, who doesn't have any emotions— whereas your robot has emotions but has firm control over them.

Kevin Emotion is a difficult topic that I'd really like to investigate, because perhaps my theory about experience being nothing other than sensorimotor contingencies could somehow be extended to emotions. What is love, for example? Do you actually *feel* love, or is love nothing more than the fact that when your loved person phones you up, you want them to carry on talking to you; or the fact that you would rather go and find them at the café than wait around reading a book in your room? Is love anything more than the assembly of all those capacities, or is there something else? The actual feel of love could be perhaps explained in the same sort of way as my sensorimotor contingencies—in which case, as a rational robot, I should manage to gain control over them.

Sue You have a lot of work ahead of you, don't you! In Tucson, Dan Wegner divided people interested in consciousness into robo-geeks and bad scientists, so I assume that you are a thorough-going robo-geek...

Kevin Absolutely.

Sue ... and you think all the rest of the people are bad scientists?

Kevin They're all robo-geeks but they don't know it.

Sue How did you come to be a robo-geek? Were you seriously thinking about these things when you were very young?

Kevin Yes, when I was ten years old my mother had this book of neuroanatomy in her bookcase and I spent hours and hours poring over the neural circuits in there. I thought this was really wondrous, but I just couldn't understand how those little neural circuits gave rise to experience.

Sue So you were grappling with the hard problem, which wasn't even called that then, even when you were a little boy?

Kevin That's right.

Sue Haven't you felt rather estranged from most people in feeling yourself to be a robot all the time, while knowing that everyone else is going around thinking that they're so much more than that?

Kevin I knew that they were all robots, and that they were just labouring under the illusion that they weren't.

Sue So is it easier, now that you are thoroughly immersed in this field, and doing these experiments and actually challenging people's views?

Kevin People are listening a bit more, but they're still very upset, because they really do feel that they're seeing everything in front of them; they really do feel that they are persons and not robots.

Sue Do you ever feel that you're detracting from people's lives?

Kevin Not at all. The fact that I'm a robot doesn't mean that I don't suffer pain, fall in love, appreciate art. It doesn't mean that I don't feel things—on the contrary, it's just a way of explaining these feelings and these experiences which doesn't necessitate any magical mechanism.

Sue Do you think consciousness can survive death?

Kevin I think in years to come we'll be able to download our personalities onto computers and have them live on in virtual worlds after we die. Then our consciousness will survive death.

Sue And do you believe you have free will?

Kevin Yes, everybody does. Even robots believe they have free will, even if they don't.

Roger Penrose

*Real understanding
is something outside
computation*

Sir Roger Penrose (b. 1931) studied in London and took his PhD in algebraic geometry at Cambridge. While there he began working on tessellation, work which led to his discovery of the Penrose chickens, two shapes that will completely tile a surface without ever repeating the pattern. He subsequently worked on many topics in pure and applied mathematics and cosmology, inventing twistor theory and working closely with Stephen Hawking among others. In 1973 he became Rouse Ball Professor of Mathematics at the University of Oxford, and in 1994 he was knighted for services to science. His work on consciousness and its links with quantum mechanics is described in his books *The Emperor's New Mind* (1989) and *Shadows of the Mind* (1994).

Sue Why is consciousness an interesting or difficult problem at all?

Roger I think there are lots of reasons, actually; one of them is the obvious reason, that there's nothing in our physical theory of what the universe is like which says anything about why some things should be conscious and others things not.

Sue Are you sure that some things are conscious and other things aren't? Wherever you draw the line, you've got a weird problem, haven't you, that you can't find out whether you're right?

Roger Well, it could be a matter of degrees; it doesn't have to be on or off.

Sue But you could never know.

Roger Oh, I see. I think that's a bit pessimistic.

Sue Is it? Why?

Roger I mean, people used to say you'll never know what the far side of the moon is like, or what materials make up stars, or all sorts of things. So I think that things which sound unanswerable at some stage, often there are indirect ways of getting at them. I mean nobody's gone into a star and taken a spoon and scooped it out; it's not like that—there are indirect ways of getting very definite information about what stars are made of.

Sue So do you think that one day we'll be able to say with some certainty that these animals are conscious and these ones aren't, or plants are, or aren't, conscious or something?

Roger Yes, I would think so, yes. But I don't think they're at all close to that now; no, it's a long way off. So don't ask me to say how I think it would be done; but that's just a general optimism I have.

Sue Now, you're a mathematician; how and why did you go from mathematics to an interest in consciousness?

Roger Well, in certain senses it's going back home, because my father was always very interested in these questions, and he took a different route: he became a professor of human genetics, and his main interest was what is it that makes people less intelligent or aware or conscious or whatever the word is; and to what extent is inheritance responsible, or environment responsible. And philosophical questions were very much his concern; as an undergraduate, he was really very concerned with these issues of what it was that made someone conscious.

Sue So you kind of grew up with these questions? And were you also asking them as an undergraduate?

Roger Yes, I did my undergraduate degree in London, and as undergraduates do, we used to talk to each other about philosophy, but more specifically, when I was a postgraduate, doing pure mathematics at Cambridge, I thought at the beginning, 'I don't have to work on what I'm supposed to be working on; OK, I'll do that too.'

There are lots and lots of lectures going on at Cambridge on fascinating things; and I went to hear Dirac talking about quantum mechanics, which wasn't my line—and that was absolutely fascinating; and Bondi on general relativity, his lectures in a completely different way were also wonderful; he was very intuitive and effusive. Dirac had none of the drama, he was absolutely precise. And another course I went to was one on logic given by a mathematician who described Turing machines, Gödel's theorem, and the various ingredients which later on I found relevant to consciousness. So these sort of side lectures I went to were all crucial to my later thinking.

Sue Why? What was it they inspired in you?

Roger Prior to that I think I would probably have been a computational-ist, looking at computers, but I had this nagging feeling about Gödel's theorem, which I'd heard a little bit about before, and I thought it was saying that there are these things that we can't know. Then when I heard this lecture it wasn't that at all: he said you can know these things, it's just that you can't know them simply by following the rules of some formal system. You have to have some method of getting at truth which is reliable, but different; you have to bring your con-sciousness, your understanding, to bear on the problem. So it's not following the rules: it's knowing why the rules work which gives you an insight beyond the rules themselves.

Sue Now, I know that your ideas about consciousness are related to the idea of non-computable functions; can you tell more of the story of how you came to that?

Roger I think already, when I was a graduate student, it became clear to me that there must be something going on in our conscious under-standing which is outside pure computation. And as I had a very scientific background, I thought: this is something that comes from outside of science; it's got to be in science in some sense, but it's not anything that you see in the science we have up to this point. And after Dirac's talk, I started to think about quantum mechanics, and it gradually became clear to me that there was a big gap in modern chemistry.

It's the sort of thing which isn't stressed in courses on quantum mechanics—Dirac was different, but if you go to a course by almost anybody else on quantum mechanics, you're supposed to get through an exam, you're not supposed to ask questions about what things mean, and what doesn't hang together: why you can use completely mutually inconsistent procedures and go away happily without wor-rying. So after a while you sort of get browbeaten into thinking, 'Well, this is what we're supposed to do; no doubt the lecturer understands this; I don't think I do.'

Sue But in some sense you weren't browbeaten were you?

Roger That's right. It's partly because I wasn't doing physics; because going to hear Dirac was a sideline.

Sue So then what happened? At some point you got into microtubules— how did that happen?

Roger Well, this was a philosophical point of view which I had: that there's something outside computation going on in our understanding, and it's probably something to do with quantum mechanics, because quantum mechanics doesn't hang together.

I had thought that I would someday write a popular book, but that was only a vague thought. And then I was watching television, with Edward Fredkin and Marvin Minsky expressing fairly extreme, hard, and strong AI points of view, and saying that maybe in the future computers will keep us as pets if we're lucky. So I thought, I don't believe this computationalist point of view, and I have a good scientific reason to not believe in it; and so this book that I vaguely thought I would write someday in the distance when I was retired—this gave it a focus.

Sue And this book was presumably *The Emperors New Mind*.

Roger Yes, so it became *The Emperors New Mind* which I wrote much earlier than I would otherwise have done, and then that evoked all sorts of criticisms, and I was totally naive and green...

Sue ... about what happens when you write a popular book!

Roger I just thought I'd write this book; I didn't even expect anybody would read it, which was the first surprise! And the second surprise was that people so misunderstood it—that was extraordinary.

Sue Well, try to explain it now to me very clearly; I think one of the arguments is that you as a mathematician can understand things or see through things that cannot be computed logically?

Roger I get attacked a little bit for that too—for saying, you know, we mathematicians can understand things that you mortals can't—which certainly wasn't my intention at all.

Sue I have this vision of you stepping off a pavement or something; am I making that up; or can you describe that?

Roger This was when I was working on a problem in relativity theory, to do with whether black holes really exist in a certain sense; I had an unusual line of thinking about it, from a pure-mathematical point of view. And a friend of mine was visiting, who's a very entertaining talker, and was engaging me in this conversation as we went across this street; now, as we crossed the street the conversation stopped—to look out for traffic; and in that period evidently an idea came to me; and then the conversation resumed at the other end, and whatever it was that had come to me was blotted out. But then when my friend had left, I had this strange elated feeling which I couldn't quite place, and I thought of all the things that had happened in the day—'Would that make me feel like this? No, I can't see that would' and so on—and eventually I got up to this point where I was crossing the street, and the idea came back to me; I thought, 'Ohhh, that was it'; and I realized that it was the key to what I needed for this problem. And the rest, although there was a lot of other work, was pretty easy after that point.

Sue So how does that relate to consciousness, because you could say that that inspiration was kind of unconscious?

Roger Well, my view on it is that in mathematics an inspiration has got to make sense. One dreams things, one has all sorts of crazy ideas which don't make any sense, but for an idea actually to take root it's got to have some sense to it. So that's where consciousness seems to me to be crucial; it's an interplay with the unconscious throwing up ideas, but in order for them to hang together you have to be able to bring your consciousness to bear on it.

Sue But you're making a very strong claim here. Here's your brilliant brain full of a lifetime of mathematical and physical ideas, and all the training and all the background and all the natural abilities—and it's all going on in parallel, masses of stuff—and at that moment something comes together. Where comes this idea that you require something more than just a computational brain doing its computational stuff? I want to understand why you feel it necessary to go beyond computation.

Roger It's the understanding which is the other side of it, which needs to be conscious, that's what I'm saying. And that's really the Gödel theory. People always attack me about this, but I think mainly they don't understand the argument; I don't know why, because it's very simple.

Sue Well, tell me, because I don't understand it. Are you saying that all this thinking and the having of the ideas and everything could be computational, but understanding isn't? Why not?

Roger Understanding requires awareness; there's consciousness involved in understanding—that's one leg of the argument.

Sue How can you defend that? How can you support that?

Roger Well, just in normal usage of the word—an entity which is not aware of something, you wouldn't normally say it understands something if it wasn't actually aware of it, would you?

Sue I don't know, I think you could say that if I did some unconscious action with a physical object that required an understanding—if I caught something that's about to fall...

Roger I would say there's no understanding there; you're doing it just through an automatic reaction.

Sue It's only a simple understanding of basic physical principles, but it is a kind of understanding.

Roger I guess you could use the word in a broader way than the way I intend it; what I mean by the use of that word is what you might call 'conscious understanding'—understanding which does require consciousness. Catching a ball you can do completely unconsciously; you're not really understanding it, you're just doing it, you see.

Sue So for you, though not for me, understanding is something more, and requires consciousness. What is consciousness, then?
I'm trying to get at the way that you're using the word, because these are strong claims, given how little we know about consciousness and how little we understand it.

Roger Well, you see this in the way that people use computers. I mean, you can use a computer to do all sorts of wonderful things, but it doesn't mean anything unless you know what it's about; what does the answer mean?

Sue So is this close to Searle's Chinese room argument and the syntax versus semantics argument? There is a difference, then, between what a computer's doing without understanding something, and what you're doing when you are really understanding something?

Roger Yes. It's not my argument, but I think Searle's argument is a valuable one; I've always thought that.

Sue So you think that Searle in his room, or Searle and the room together, wouldn't understand Chinese?

Roger That's right, yes. I agree with him on that.

Now, let me finish the Gödel thing, because this is the crucial argument. It's also the one I get in most trouble with, I think because it is a very strong argument and people then try and find holes in it.

The argument is—I'll put it in simple terms—suppose you're trying to ascertain the truth of clear-cut mathematical statements like Fermat's last theorem, which says, roughly speaking, that there's a certain computation that wouldn't ever stop. Now, for those very simple types of mathematical statement, there is no argument about which ones are true and which ones are false; you might have to work hard to see, but the fact that it's an objective thing whether they're true or false is not controversial. Now, how do you come to the realization that certain of these things are actually true? Well, you might use some kind of axiomatic system, you might have some rules or procedure, which, if you apply these rules correctly then you must trust the conclusion. And what Gödel showed is that any such system of rules, provided it's not too trivial, will have the property that your belief that those rules only give you truths enables you to transcend the rules. So you can, if you like, state that the rules are consistent: if the rules only give you truths they must be consistent; and if they must be consistent then that statement which asserts their consistency, which is another statement of this kind, lies outside the scope of the rules themselves. So how do you know that those things are true, that you don't obtain using the rules? Well, how do you know anything that you obtain using the rules is true, you see? But you can trust the rules only if your understanding tells you these rules are good rules, that they won't give you nonsense.

That understanding which tells you they don't give you nonsense gives you something beyond the rules. So it's the understanding that is not constrained by any system of rules, because you make those rules try to imitate what the understanding is doing, and your understanding immediately leaps outside it.

Sue And this kind of understanding, is it uniquely human?

Roger No, I don't think so. Of course, somebody could say, 'Well, my dog doesn't understand Gödel's theorem'; but somebody could come up to me and say, 'Well, I don't understand Gödel's theorem'—and that doesn't make him non-human, does it, or not conscious? And the fact that his dog doesn't understand Gödel's theorem, likewise, it doesn't make *it* non-conscious.

Sue But I'm trying to get at the principle here, of what kind of a thing potentially might have this special understanding that you've described.

Roger Well, I think that understanding generally does that. An entity would not in my mind be considered to be intelligent if it has no understanding.

Sue So could you build potentially, in the future, a fantastic robot to have that sort of understanding?

Roger Well, if a robot means a computationally-controlled system...

Sue I do mean that; I mean a computationally-controlled system.

Roger Then I would say no, it will never be intelligent; it can play chess very well...

Sue ... but not be intelligent in the sense of understanding in the way that you're using it? So what is required, then?

Roger It's very difficult to tell. It's not that hard to tell in practice, but you can see in these Turing tests—every year they have these huge competitions, and they're still pretty stupid. You'd think with this fantastic computational power that these machines have, way way way beyond what we can do, computationally—and still they're stupid.

Sue Many people think they're getting better and better and it's only a matter of time until they really understand things; now you're saying there's something special about understanding that they haven't got and they won't ever have?

Roger You might be able to imitate it to a degree, but it won't be the real thing.

Sue OK, tell me, what's the real thing?

Roger Well, the real thing involves awareness. You see, I think the Gödel argument is actually extremely rigorous, although most people attack me and say, 'Well, Roger's arguments, they're interesting, but they're basically flawed.' I say, 'Oh, well tell me where the mistake is.' Nobody has done that. I've waited. They can be rude to me, but they haven't pointed out a mistake.

Sue Well, I'm in no position to point out a mistake, but I can point out this leap that you're making—saying there's something extra that is 'real understanding'—and I want to know what it is.

Roger Yes, well then it gets more conjectural and I admit that.

Sue OK, conjecture away.

Roger I'm saying that the Gödel argument tells us that we are not simply computational entities; that our understanding is something outside computation. It doesn't tell us it's something unphysical, but there's a crucial thing that's missing, which has to do with quantum mechanics. Mine is a version of the Sherlock Holmes argument, which I admit is a weak argument—that to say once you've eliminated everything else, then what remains must be the truth, no matter how improbable. Quantum mechanics is the most obvious place where we don't know enough about physics. Where do you see non-computability in physics? You don't seem to see it anywhere else. So this, therefore, is presumably where it is.

Sue So you arrive at this idea that consciousness requires some kind of quantum computation.

Roger It's already unconventional to say that the workings of the brain require quantum mechanics in a fundamental sense. But even that's not enough, you see, because I require going beyond standard quantum mechanics, I require something which involves an improvement on quantum mechanics, now that is unconventional even in the quantum world.

Sue Well, let's say we take these two unconventional steps; now where does that get us with understanding what the brain is doing and where consciousness comes into it?

Roger What I'm trying to say is that if you need the brain to do non-computational things, you've got to find something in the brain which has a reasonable chance of isolating large-scale quantum effects, and that's where the microtubules come in, and this I got from Stuart Hameroff.

Lots of people write to me often, with completely crazy theories; so I get this letter from somebody I've never heard of before, who's talking about these funny little tubes which lie in cells, and I think, 'Oh, not another one.' But then there are pictures of these things, and this is just what I want, because nerves disturb the environment far too much. There's no chance, in ordinary nerve propagation, of shielding the signal from the environment. Now, with microtubules, it looks as though there's a real chance.

Sue Doesn't it bother you that, firstly, there are microtubules in every cell of the body, and secondly, they're generally thought to have a structural function that explains why they're there and what they do?

Roger Let me answer question number two first. I use an analogy here. We already know what noses are for—noses are for filtering the air, and smelling things, and so on—now, when we see an elephant, what does it do? It uses its nose for everything, for washing, and picking things up, and building things. Just because we know one of the major things that microtubules do simply doesn't tell us that they don't do something else in appropriate circumstances.

Sue And what about point number one, that they're in all the cells and not just in the brain?

Roger There are different possible answers to this question. My guess has to do with the A and B lattice structures. There are two different structures that have been suggested for microtubules: the original one, which I described in *Shadows of the Mind*, is the A lattice, and these are the nice symmetrical ones. The B lattice looks very similar to the A lattice but they are unstable and keep coming apart. And all the arguments that people produce that they can't be conscious refer to the B lattice.

It's also apparently true that in neurons you get stable microtubules: the ones that somehow disassociate and come apart and come back together again are what you normally find in cells, but you get stable ones in the brain.

Sue As I understand it, Stuart originally had this idea because he saw the effect of some anaesthetics on microtubules and that's why he thought they might be involved in abolishing consciousness; and then he subsequently discovered that most anaesthetics don't affect the microtubules at all; so the original purpose of the theory was thrown out. Do you still stand by that theory—do you still think the answer might be there?

Roger I'm certainly open about microtubules; I think it's only part of the story—that is my guess. People often say these hypotheses are untestable, but there are lots of ways you can look to test this kind of hypothesis, and there's all sorts of circumstantial evidence too, which I've never heard being discussed. There are also these nanotubes which are a bit like microtubules: they're much smaller and much more evidently quantum-mechanical entities—and you can make these nanotubes so that they twist one way or the other way. Now this is analogy. I'm not saying microtubules have this property but it's circumstantial evidence.

Sue If we look to the future, I imagine that many of the people I've talked to in consciousness studies, particularly functionalists and identity-theorists, would say what's going to happen is that we'll just learn more about computations in the brain, more about perception, more about learning, more about memory—and the hard problem of consciousness will simply go away.

Presumably you are saying something very different—that what will happen in the future is we'll learn more about all these different chemical and physical structures, and eventually—wow—we'll see this entirely new process and that will be what explains human understanding and human consciousness?

Roger Yes, that's very fair. But I would say that our understanding of the physical world is much more limited than people think. Physicists are usually very arrogant people so they'll claim almost everything; but my view is that there's this physical world out there that we know an awful lot about, but that there are big things we don't understand yet; and I'm claiming that non-computationalism is one, but it's for most purposes a tiny minor thing which you don't even notice.

Sue But it's a major thing if it explains the great mystery of what it means to be conscious.

Roger Exactly, exactly. So it's sort of lying in wait, and only when evolution has got to a stage where it can actually start to latch into this and make use of it—that's when you start to get consciousness.

Sue I want to ask you about the philosopher's zombie—the idea that you could be Roger Penrose, sitting here doing everything you do, speaking as you do, but it all be dark inside. On your understanding of consciousness, would such a zombie be possible?

Roger A philosopher's zombie is something which I would say couldn't exist. I'm more of a functionalist than that, but I'm not a computational functionalist.

Sue I know that Stuart Hameroff thinks that these quantum coherent processes could survive the death of the brain that created them, and that therefore there could be life after death; do you agree with him about that?

Roger   I would have a lot of trouble following that one. I'd say there are lots of things we don't know, and so I would hate to be dogmatic about things of this nature. But I certainly don't see that yet.

Sue   **Do you think you have free will?**

Roger   Free will is a really deep, difficult problem. When I talk about non-computational things, that's not free will, because you can have deterministic systems that are not computational. But then that's never enough, and the Gödelian argument says you must have non-computational things. You keep applying it again and again—so even if you have what Turing would call an Oracle machine, which isn't computational but it's the next step up, there again you can Gödelize that and say, no, no, that can't be it either. So you keep going, step by step; this is one of these unending chains of argument—maybe the argument doesn't work well beyond a certain point, but I don't quite see why.

One actually enters parts of mathematics where even mathematics isn't fully understood; you have issues which probe the borderline of our understanding of mathematics; so I think there's something very subtle going on. I think free will doesn't even mean what people normally think it means. For instance, free will is usually talked about in connection with moral issues: do you have freedom to do this or that? And I suspect it's tied up with these very profound issues. So I would say, in answer to your question do I believe in free will, the simple answer is I don't know.

Sue   **Does that moral element come in for you when you're thinking about the nature of consciousness?**

Roger   I think it has everything to do with it, because without consciousness somehow morals evaporate. I remember having an argument with a computationalist—I've forgotten who it was now—where the question of morality came up, and this person just didn't understand what consciousness had to do with morality, and I thought, 'What?!' I mean, if you bought a computer which is conscious you'd have a responsibility; it's a moral issue.

I remember talking to some people about space travel, and they were interested in making computer-controlled devices that went and sat on planets, and if they were really intelligent then they didn't need to send people. But if they're really intelligent maybe they have to be conscious too—they said, 'Yes, yes, well of course they'll be'; but then if they're conscious, well, you've got to bring them back, you've got a moral responsibility to them.

Sue   And I suppose, if you're right about the nature of consciousness, then this new physical understanding will help us with these moral issues.

Roger   Maybe ultimately, but I think it's a long way off. Microtubules might not be the answer for all I know. I don't mind being attacked by people as long as they're rational about it; so many people seem to be totally off—not just irrational, but they don't understand my point of view; they accuse me of having different points of view from the one I have.

Sue   But don't we just have to put up with that, as investigators of these difficult things?

Roger   I suppose so, yes. But at least, I think, the study of consciousness has become more scientifically acceptable. Not that many long years ago, you know one lived it in secret. Now it's something you can do and not be thought to be a complete crackpot.

Vilayanur Ramachandran

*You're part of
Shiva's dance; not a
little soul that's going
to be extinguished*

Originally from India, Rama (b. 1951) trained as a physician in the USA, and then did his PhD at Trinity College, Cambridge. His earliest research was in vision but he is best known now for his work on neurology and synaesthesia, as well as his interest in Indian art and the connections between art, vision, and the brain. He is Professor of Neurosciences and Psychology, and Director of the Center for Brain and Cognition at the University of California, San Diego, and Adjunct Professor of Biology at the Salk Institute. He is author of *Phantoms in the Brain* (1998) and *A Brief Tour of Human Consciousness* (2004)

Sue What's the problem; why is consciousness so interesting and difficult?

Rama It's the biggest challenge to science, because all the problems we have tackled and solved so far have to do with the external world, such as DNA, or the Earth not being the centre of the universe, or cosmology, or string theory. But we're now finally confronted with in some ways the biggest problem of all, namely, understanding the very organ that made all those other discoveries possible, turning on itself and asking, 'Who am I?'

I don't mean this in the metaphysical sense—simply, here's this blood and flesh generating this amazing sense of self, and questioning its own origins and its own future; what happens when I die?—all those questions which people have been preoccupied with for thousands of years we can now begin to approach scientifically.

I'm always confronted with this because I see patients who have brain damage which changes your sense of embodiment, your sense of self, your qualia, all of these things that philosophers discuss. I'm dealing with this empirically every day you know.

Sue You said two highly contentious things there: the first one is that you distinguish between an internal world and an external world; and the second one is that you threw in qualia. Let's come back to the qualia and tackle the other first—do you really think there are two worlds? What do you mean by an inner world and an outer world?

Rama Well, instead of doing that, let me just state my position. I think that people have falsely dichotomized or separated the qualia problem from the self problem. I think that they are two sides of a coin, for lack of a better metaphor. In other words, if there were no such thing as the self, then there'd be no qualia—because, to put it very crudely, you wouldn't know. You can't have free-floating qualia without an observer who experiences the qualia, so the concept of self is implicit in the concept of qualia. Nor can you have a self without any qualia—any emotions or bodily sensations of any kind.

Even though Eastern mystics are certain that you can; they say that you can go into an isolation tank and then pretend you're completely unconscious, not have any qualia, but that the sense of self still endures; that therefore the self can exist independent of qualia, indeed independent of the body.

Sue But Zen practitioners say exactly the opposite: that the qualia—well, they wouldn't use that word—the arising phenomena don't disappear; the self who is experiencing them disappears or the two become the same thing. When that happens there's just experience without a self. You're saying that's impossible, are you?

Rama I'd say that's impossible. I think that the two are logically two aspects of one phenomenon. It's a bit like the Möbius strip: both sides have to coexist.

Sue That doesn't mean it's impossible, though, because then you could say, using the Möbius strip, that it would be as though one were able to see both sides at once. Might that be one interpretation of that experience?

Rama Like all analogies, one can only push it so far. On local inspection it looks like two different phenomena, like two different surfaces of the Möbius strip, but in fact there is a coherent scheme you can come up with, in which they both form part of one continuous reality. Now, that's an analogy, OK? Now let me get more specific. I think what's going on is—let me make some bold assertions—first of all I think animals don't have consciousness or qualia.

Sue None of them! Only humans, right?

Rama Great apes come close. I think there is a quantum leap. There is something very unique and special about humans, not in any theological or mystical sense, but just in terms of functions.

Sue You mean it's not because they have a soul, but it's something to do with a function. But what function? Some people would say language, some a sense of self; where would you put that special leap?

Rama I think those two are related, by the way. But let me first assert, and then I'll give you the evidence as we go along, that lower animals—I'm not supposed to call them lower animals, but animals in general, even higher primates, excluding humans—have only a raw background awareness. But they're lacking extra stuff which I have called meta-awareness.

Now, this could be another parasitic brain, to put it crudely, that uses the output of the 'first' brain as its input. In other words you first have processing of information, and various automatisms of the kind done by the dorsal stream, and then some stage in evolution created a representation of the representation for other purposes. The question is, what are those other purposes?

You could say, isn't it redundant; why create another representation of the representation? The answer is, it isn't redundant; you're doing it to fulfil a new computational need, namely open-ended symbol manipulation in your head. This is what we call thinking: coming up with outlandish conjectures which are made by the imagination, by juggling these symbols in your head. And closely linked to that is the emergence of language: being able to communicate these ideas, intentions, and thoughts with other people; and constructing a theory of other minds. All of this happened more or less simultaneously in evolution, and it was a quantum jump in the mind of an ape.

Sue Now, before we go any further I want to take you up on this whole business of qualia. You're using the concept here all the time, so first, what is a quale?

Rama Well, there are various ways of stating it.

Sue No, I don't want to hear all the ways of stating it. For you, when you're talking to me about this evolutionary jump and how qualia suddenly appeared, what exactly are you talking about?

Rama OK. The only way I can state it clearly is with the old famous thought experiment. That is, let's say you're a Martian super-scientist, and you are colour blind. You come to me and you say, 'Rama, I want to figure out how you see long wavelength, and I'm going to look through your brain at all the patterns of activity.' You get to my Broca's area and you say, 'Red: the muscles are active,' and you say 'I think of red apples: memories are active.' All of this happens, and you think that's all fine. But this description does not contain my ineffable experience of redness which I can never tell you, because you're a colour-blind Martian.

Sue So would you fall into the category of people who think that as well as brain activity there's something extra: the qualia, the experience, the subjectivity?

Rama Well, you need to explain how it comes about. I'm not saying there's any spooky stuff going on.

Sue All right, then do you think that if you understood all of the information processing, the spoken language—all the things that the brain was doing—then you would understand all of the experiences?

Rama I think you would understand qualia, in the same sense that you understand an electron, or anything else; you don't then go on to say, 'What the hell is it? There's something ineffable you can't communicate about an electron.' You say, 'This is it'—and you're going to be able to do that with qualia as well.

But I'm saying the description in mechanistic terms by the Martian scientist does not contain within it the experience of qualia. To explain that you need to take an extra step; and I think that extra step is that at some point in evolution the sense of self emerged, and that requires this meta-representation.

Sue In some of your writings you've used phrases like, 'Some neurons are qualia-laden', or 'Some have qualia attached to them', or something...

Rama That's just shorthand. What I'm saying is that these circuits are qualia-laden. Someone could assert that the spinal cord in itself experiences qualia; but that's misuse of the word. I'm saying that qualia cannot exist without the self which the spinal cord lacks, by definition, so therefore that is a misuse of the term qualia.

Sue So you shouldn't say it then, should you? But let me get this straight; see if I've understood it. When you say something is 'qualia-laden' I infer as a reader that you mean somehow these separate things called qualia are sort of attached to these neurons...

Rama No, no, no, no.

Sue So you're actually denying that you mean that at all. Oh good.

Rama I'm not a dualist; I'm a neutral monist, but the trouble with neutral monism is that it doesn't go far enough; it doesn't say exactly what's going on. So I'm trying to push it to the neural circuitry and say that once the sense of self emerged ... See, it's kind of a funny problem, because in a sense you have to know that you know, otherwise you don't know. That's the crux of the matter, and that's why you need the sense of self, which knows that it knows, or knows that it is seeing red.

But it's not an endless regress. I can say, 'You know that I know that I had an affair with your wife'; but if I say, 'I know that you know that I know that you know that I know,' you start losing the thread, like an echo. There are only so many steps that the brain can handle, and that's adequate for the sense of self. So it's not an endless regress, it's other brain structures and there is no homunculus.

Sue Is there evidence for all this?

Rama You can look at all the brain-lesion studies from this point of view, such as Weiskrantz's blindsight phenomenon. Here you have a sensory representation, but you don't have a meta-representation; or it's uncoupled from that. That's why the chap moves his finger and touches something but is unaware of what's going on. His 'self' is uncoupled from it, and he's unaware of what's going on.

Conversely, in Anton's syndrome, when a person is completely blind because of damage to the visual cortex but says 'Well, I can see fine'; but if you ask him to touch something he can't; he has a spurious meta-representation.

You can talk about hypnosis, or about every conceivable clinical neurological phenomenon that affects consciousness, from this point of view—this dichotomy between having the representation versus having the representation of a representation.

Sue What about pain? You've made a case here that qualia come about only with this big leap to self-concept and language. Now, when my cat comes limping in through the cat flap looking pathetic with a thorn in his paw and I have to take it out, and then he seems happier, I think he is experiencing pain, in the sense that he's got the painfulness of pain—he really doesn't like it; it feels like something to that cat. Of course I can't know, and I can be as sceptical as anything about it; but that looks to me from the outside as being as much a claim for qualia as it would be if I poked you now and hurt your arm.

Rama I know what you mean, but I think that's not the case. I think that, for example, your withdrawal from a hot kettle is a different pain from the pain that you then contemplate. In the first case, the pain of withdrawal from a kettle, there is no qualia, and no meta-representation. In the second case, when you contemplate the pain, you have a meta-representation, which you can communicate with others; it has all kinds of links with memories where you say, 'Oh, pain, that's a bad thing; let me not do this again; let me tell this other chap about my pain; let me take some medicine for this pain.' It's got all these vast semantic implications; and you need those and the link to the sense of self, in order for fully-fledged qualia to emerge.

I think the cat is responding to the pain with a reflex withdrawal. So however much you might be tempted to infer that it's contemplating its agony, it isn't. Similarly, you could say that if somebody's under anaesthesia, you've uncoupled the person, the self, and therefore qualia as experienced by the self, from the pain. Someone could argue 'How do you know the spinal cord isn't independently conscious on its own?'; so if you do a spinal block, is that unethical? That's no more a problem than the cat problem.

Sue But morally the cat problem is serious because of factory farming, because of all kinds of cruel practices to wildlife...

Rama But these are cruel practices to the spinal column...

Sue Yes, but what is your answer about the factory farm? Does it matter? Do you want people to treat the animals better, or does it not matter because they're not having qualia?

Rama To me that problem is like abortion; in other words, you're confounding ethics with science. Somebody will always make the case that you're preventing a human being from existing. It's also a bit like asking 'Is a virus really alive?' In the post DNA era—now that we know what a virus *is*—it's no longer useful to ask 'But is it really alive?'

Sue That's a different argument; let's stick with the consciousness argument—I'm really not going to let you get out of this; you're going to stick with consciousness! So let's follow your theory through logically, let's say a cow is going to the slaughter; you can kill it instantly, or you can kill it in a slow way which to us would be very painful. Now, do you think it matters?

Rama I wouldn't say that. It doesn't experience pain like we do; it certainly can't introspect on its pain. It's a bit like that 'Is the virus *really* alive?' problem again; we don't want to get distracted by semantics. I think that as mammals we empathize with certain behaviour patterns, and this makes you think that the cow has qualia, and therefore you shouldn't hurt it. But then you can say, 'Well, why am I vegetarian?', you know; once you start getting into ethics and start asking at what point does an embryo become conscious and therefore you can talk about murder versus just abortion...

Sue So you're saying something like this: 'I'm a vegetarian, I don't want to eat animals, I would rather they were killed in a nice way, but actually I don't think they feel pain.'

Rama That's correct, I would say that, if pushed.

Sue Fair enough; you are being pushed; you have been pushed!
Now, I want to change tack completely; how did you get into all this in the first place? There you were, trained as a doctor, then what?

Rama Well, if you are trained to be a physician and you're examining neurological patients, it's inevitable that you become interested in consciousness. Seeing people with strange mental phenomena forces you to confront this problem.

Sue Oh, but an awful lot of neurologists just stay away from consciousness; perhaps they think it's dangerous scientifically; but you're one of the unusual people who's prepared to get tangled up with it. What do you think made you different?

Rama I think partly my training back in India, in the early days. People often get brainwashed by the scientific community; behaviourism had a pernicious influence, and people said it's not fashionable to think about internal mental states. People also said this about vision, you know—you're not allowed to ask the subject what he's actually experiencing. It was Richard Gregory who turned the tide in some ways, and revived the Helmholtzian view; and I think I've done that partly for neurology. There was a golden age of neurology when everyone in neurology was interested in this.

Sue Yes—there was Hughlings Jackson...

Rama ...and Charcot, Freud, all these people. And then it was eclipsed by behaviourism; in neurology they said, 'Don't ask the patient what he's experiencing because you'll be misled,' and that of course threw the baby out with the bathwater. But I think that in this generation I have done a lot to try and revive that approach of good old-fashioned nineteenth-century neurology—because I was untrammelled by fashion.

Sue So it was coming from India and being right outside this climate that enabled you to say, 'I'm not going to fall for that memetic indoctrination.'

Rama Yes.

Sue And what was the first research that you ever did that was directly related to consciousness?

Rama My very first experiment, which was published in *Nature* when I was 20 years old, was in a sense about qualia and consciousness. I had a stereogram, and put vertical stripes in one eye, and horizontal stripes in the other eye. The amazing thing is, it was still seen in full stereo even though you're only seeing one eye's picture at a time, because of the rivalry; so I said, look, the stereo mechanism can extract its disparity information without being conscious of one eye's image; so already there's this distinction between non-qualia and qualia.

Then I came across some of Richard's well known experiments; and mainly worked on psycho-physics and perception. And by the way, in those days that was unfashionable too. Richard was doing it, and Bela Julesz, but really there was nobody else.

Sue You are absolutely in the Richard Gregory mould aren't you? I can see why he must have been such an inspiration to you. But did you get any flak from other scientists or neurologists for dabbling in such a topic as consciousness?

Rama In the beginning I did, but not now. I think that scientists are always more receptive if you get things right. People sometimes come and say, 'Isn't this controversial?' and I say, 'Well, I've got 35 years of publication; show me one empirical finding which somebody has questioned.' There isn't one.

Even the very speculative ideas, for the most part have stood the test of time. So then people are more receptive to your more speculative ideas. What I'm saying is you need to pay your dues, and so long as you're doing that, then in parallel you can say outlandish things, and be speculative—like when I'm talking about meta-representation, it's only speculation, but people are more forgiving of it.

For example, lots of people have confirmed the work we've done on phantom limbs; and then if you start talking about qualia, people listen to you.

Sue I haven't done as wonderfully dramatic demonstrations as you, but I think I've had exactly the same experience through my life, that people will listen to some of my mad speculations because I spent 30 years doing experiments.

Rama Exactly.

Sue I want now to go back to the central issue, the whole qualia thing. Do you believe zombies are possible?

Rama No, they're not possible. I think if you create a creature which is identical to us—it doesn't matter how you create the zombie—it'll be fully conscious in the human sense.

Sue Does it have to be physically identical or could it be identical functionally?

Rama You mean if it's made of dinner plates or silicon chips? I'm agnostic about that; I think that it's the information flow that's critical, so in that sense I'm a functionalist; but I'm not sure.

Sue Do you believe in free will?

Rama Well, I think that once this meta-representation evolved, that for some reason had to be linked with the sense of volition. And I can tell you why, again in terms of neurology. I recently suggested a variant of the famous Libet experiments; I believe Grey Walter did something similar, although he never published it. You know that in Libet's experiment, if you wiggle your finger you get a readiness potential, and it turns out you can tell somebody, 'In the next ten minutes wiggle your finger three times, anytime you will it.' Now you do the experiment and you find that the readiness potential actually happens half a second to a second earlier; so this has a paradoxical flavour to it, but in fact it's not a great paradox: there is some internal sense of willing. The idea is, how come the sense of willing...

Sue ...comes too late? The traditional interpretation of Libet's experiment is that the sensation of willing comes after the beginning of the readiness potential, so it can't be causal. Do you have a different interpretation?

Rama No, no, I have a Dennett-like approach to this; I'm saying that there's a spatio-temporal smearing of events in the brain. But it'd be lovely to try this following experiment: you take the readiness potential and give the person feedback on the computer, and say 'abort', 'stop', or 'move your finger'. What'll happen is one of three possibilities. The person will say, 'Oh my God, I don't feel free will anymore, I do everything the computer's telling me.' Or he'll confabulate, and say, 'Oh no no no, I thought of it first,' and rewrite the time sequence. And the third possibility is that the machine has precognition—well, it only feels like it; in other words, it may make him relinquish his sense of will to the machine—'The machine is bloody controlling me'—as paranoid schizophrenics do.

Sue Can you do it on a single trial?

Rama We're trying to. It's extremely hard to get an EEG signal in single trials, which is what we need; or MEG is a possibility. We're trying to get a good signal with EEG.

Sue And what about free will?

Rama Free will. So what is required is to create a meta-representation of a volitional action. In other words, you create a representation of your intention and your desire to perform the action, which comes in the anterior cingulate, along with the limbic structures. So you need to desire and to anticipate, and then you need to decide, and then you call it a volitional action, OK? If that's uncoupled, then the subject has apraxia. It's a classic example; it's all about free will caused by an uncoupling of the meta-representation from the representation. So an animal has a representation of the action, but it does not have a meta-representation, which is unique to humans with the emergence of sophisticated new circuits in the supramarginal gyrus and anterior cingulate.

Sue Dan Wegner says that free will is an illusion created in three steps: he says that first you have a thought about an action, then the action happens, and then you conclude that the thought caused the action, when actually something else, some underlying processes, caused them both. Yours is a more complex scheme, but would you essentially agree with him?

Rama That's not saying much more than that it's spatio-temporal smearing, à la Dennett. It's a post hoc rationalization.

Sue Yes exactly; it's a kind of confabulation. So that if you believe that your conscious thoughts caused the actions to happen, you're wrong. Would you agree with that?

Rama I would agree with it, but what I'm arguing is that you need to go a step further and talk about which brain structures are involved and what is the nature of the representation. The desire component comes from the anterior cingulate, and the anticipation component is a meta-representation which comes from the supramarginal gyrus.

Sue Thinking that about free will, how does it affect the way you act in the world?

Rama I think it's a bit like the whole dance of Shiva thing, that you think you're an aloof spectator watching the universe, but actually you're just part of the cosmic ebb and flow of the world; but it doesn't change anything.

Sue Doesn't it change anything? I mean, that's a wonderful way of describing it, the dance of Shiva, and realizing through your science that you're just a part of this grand dance.

Rama It's ennobling, rather than diminishing. It's only when you start thinking that you are some aloof thing which is in charge of everything, that you become scared of dying, because you say, 'Oh my God, when I'm dead, I'm not around anymore.' But if you think you're part of the ebb and flow of the cosmos, and there's no separate little soul, inspecting the world, that's going to be extinguished—then it's ennobling. You're part of this grand scheme of things.

Sue I absolutely agree with you.

Rama Dawkins would be annoyed with us, because he'd say I'm trying to let God in through the back door, to use his expression. But it's not through the back door; I think it's a perfectly legitimate view.

Sue But, from what you've said, you realize that many people find this idea really distressing, and they can't make that leap. Do you think that all the science that you've done, and your thinking about consciousness and so on, helps you to make that leap? Or is it that you were brought up in India, understanding Hinduism and concepts like the dance of Shiva?

Rama The science helps me make that leap, I would say. And I think it's silly when, you know, somebody comes and tells you, 'Here are the neural circuits when you're having an orgasm,' to then say, 'My God, that's all there is!' But the kind of world view that comes from Hinduism, I think that was a peripheral thing; I don't think it affected my main research in any way, except to the extent that maybe you're more interested in the phenomena of consciousness if you come from an Eastern tradition.

Sue Do you do any first person practice of any kind; do you meditate, for example?

Rama No, I don't, and I'm ashamed of that, because, you know, everybody asks me that. I have an open mind about it; but I think sometimes there's a little bit of what Freud would call reaction formation: you come from a tradition and you deliberately stay away from it, because you say, 'What has this really resulted in?' But now I'm much more open to such ideas, and it's worth exploring scientifically. The trouble is, the people who study these things are very often on the fringe; and the studies are often not properly controlled.

Sue But do you think that having first person experience of meditating, or of anything like mystical experience, would help you as a scientist?

Rama Almost certainly. What you need is people who are willing—and you've done it to some extent—to develop links between these two domains of enquiry, so that Eastern mysticism becomes legitimate or not legitimate. You can find out.

John Searle

I don't understand a word of Chinese

John Searle (b. 1932) is Mills Professor of Philosophy at Berkeley, where he has been since 1959. He says he is, and always has been, 'interested in everything'. After studying at the University of Wisconsin, he spent three years as a Rhodes Scholar at Oxford and became a don at Christ Church College. He has won numerous awards, and whole conferences have been devoted to his work. His Chinese room thought experiment is probably the best known argument against the possibility of 'Strong AI'; a term that he invented. He says that brains cause minds and argues for biological naturalism. He has written books on language, rationality, and consciousness, including *The Rediscovery of the Mind* (1992), *The Mystery of Consciousness* (1997), and *Mind: A brief introduction* (2004).

Sue There's something special about consciousness, isn't there, that caused it to have been expelled from psychology for so long, and that brings us to have a whole new field about it? What is it that's special about consciousness?

John In a word, consciousness is our life. If you think about the sequence of our life, the things that matter to us after birth and before death are forms of consciousness, and so the funny thing is not, why is consciousness important, but, how can anything else be important?

And the answer is, of course, that other things are important in relation to consciousness. We're happy if we make money because then experiences are possible that we wouldn't have otherwise; we're depressed if we're under a totalitarian regime because then the form of our conscious life is made miserable, and so on. So what's special about consciousness is that as far as human life is concerned it is pretty much the precondition of everything important.

Sue **But you talk there about consciousness as important as compared with other things—but that raises the whole question: are there other things?**

John There are other things, sure: there's digestion, there's photosynthesis ... When I say consciousness is all-important, I don't want to even hint at idealism that suggests that all of reality is just forms of consciousness—I don't believe that for a moment. Consciousness is an amazing product of certain kinds of human and animal brains, but it's very local, very special.

Sue **But do you think it is somehow different in kind from all those other things?**

John Yes, absolutely. I'm sorry, I didn't understand that was the question you were asking me. What's different in kind is this: consciousness only exists as experienced, or enjoyed, by human or animal agents, by some conscious beast, some 'I'. I like to put that by saying it has a first person ontology, whereas mountains and molecules and tectonic plates have an objective ontology, a third person ontology—they're just there.

Now, a lot of people mistakenly think that means you can't have an objective science about consciousness, but of course you can; you can have an epistemically objective science about a domain that's ontologically subjective; that's just a fancy way of saying that in your account of knowledge you can get objective knowledge about a subjective domain.

Sue **But you're getting here at the hard problem, and I'd love to know what you think about that.**

John You mean the problem of how consciousness fits in with the brain—the mind-body problem, as it used to be called. Well, I think philosophically it has a rather easy solution; the hard hard problem is neurobiological. So, I'll give you the easy solution first. The easy solution is: look, we know it's all caused by brain processes;

all of our conscious states, every damn one of them is caused by neurobiological processes in the brain—and the word there that counts is 'cause'. The brain is a biological organ and, like other biological organs, it's a causal mechanism, and it functions to cause conscious states and processes. What are those states and processes? Well, they have this subjective, qualitative feel, but they exist as processes in the brain; they are higher-level features of the brain. So you can summarize the relation between consciousness and the brain by saying, first, brain processes cause consciousness; lower-level neurobiological processes cause conscious states; and second, these conscious states are themselves higher-level features of the whole brain system. So it's a bunch of neural firings that cause the conscious state, but the conscious state is not identified with any particular neuron: you can't pick one out and say, this one is thinking about your grandmother.

In one sentence: consciousness is caused by brain processes realized in the brain system.

Sue **But I still have this lurking feeling that there's something that doesn't quite work here; that consciousness, or subjectivity, or what it's like to be feeling like this, me, now—how can that arise from something objective when they seem to be such totally different kinds of thing?**

John   That's a wonderful expression of the traditional mind-body problem, and that's where we get to what I think is the really interesting problem: how does the mechanism work; how exactly does the brain mechanism produce this? However, we have to be very careful, because there are two ways of hearing your question. One way is hearing it as asking, 'Look, we know it happens, let's get into the plumbing and figure out where and how it happens'; but there's another tone of voice in which that question is asked, which suggests we can never know, it will forever remain a mystery.

I don't believe that second line; I think the first line is right; we know it happens, we know the damn brain does it. Here we have three pounds of this gooky stuff in our skull, a kilogram and a half; and we know some processes in there are causing consciousness. We start with that fact, we take it as a given, now then let's figure out exactly how it works.

I have to say, I think not all but most neurobiological research is based on a deep philosophical error: it's what I call the building block approach. What these guys try to do is to find the neural correlate of individual conscious phenomena; and the idea is that if you could find even one building block—such as what it is that causes me to experience red—then you could crack open the whole system.

That might turn out to be right, but I think it's a mistake. I think you have to take seriously the idea that consciousness, as created by the brain, is a unified conscious field and that what we think of as perception doesn't so much create conscious states as modify the pre-existing conscious field. So the key question is not, what is the corre-late of each particular conscious feature—such as the taste of beer or perception of the colour red—but rather, what is the difference between the conscious brain and the unconscious brain? That's much harder than the building block approach because you have to look at massive sequences of synchronized neuron-firings over large areas of the thalamo-cortical system—you've got to deal with big chunks of the brain.

Now, I said they were making a philosophical mistake, but of course this won't be settled by philosophy; it'll be settled by actual neuro-biological research; it's a scientific and not a philosophical question, and maybe I'll be proven wrong.

Sue You talked about working on what some people would call the easy problems—trying to understand how the system does it, as opposed to giving up completely and just saying 'It's a mystery.' But there are other people who say we need a fundamental new principle in the world, or we need quantum mechanics. What do you think of those possibilities?

John I'm all for trying anything, it's just that on any given day, when I get up and go to work, I have to go to work on the basis of what we know now. What we know now suggests that you'd better take the neuron and the synapse seriously. Maybe it'll turn out that we're wast-ing our time on all these dumb neurons, and you've got to get inside there to the microtubules, down much lower than the pathetic neu-ron and the synaptic cleft; or maybe you've got to look at much big-ger things than neurons, you have to look at whole clouds of millions of neurons operating in chaotic dynamics.

Maybe we're going to need some quantum mechanical explanation, but I'm suspicious because most of the quantum mechanical accounts I have seen of consciousness are obviously not going anywhere: they substitute two mysteries for one. Consciousness is a mystery; how're you going to solve it? Oh well, here's another mystery, quantum me-chanics. So now we've got two mysteries, but I don't see that we've got a solution to either. However, I don't want to give you the impression that I'm opposed to that research; let a million flowers bloom; let all these people try these research projects.

Sue Could you explain something more about your conscious-field theory?

John I like to think of it this way: if you wake up in an absolutely quiet dark room, you can become fully awake, fully conscious, even though you are receiving absolutely minimal perceptual stimuli; if you think about it you can feel the weight of your body against the bed, and the weight of the covers against your body, but otherwise you have no perceptual stimulus at all, but you're fully conscious. What I want to know initially is, what's the difference between your brain now, and your brain five minutes ago? The problem is that, with current scanning techniques, the conscious brain looks a lot like the unconscious brain.

OK, now then here I am, lying in bed in this dark room; now I get up and move around, brush my teeth, turn on the lights or open the window or whatever; I have a lot of experiences I didn't have before, but the claim that I am advancing is that we should think of those not as new creations of consciousness, but as modifications of the conscious field that began when I woke up. So if we were following my research project, I would suggest that our best bet is not to go for the neural correlate of some specific sensory mode, such as the experience of the colour red, or of the sound of middle C; but to try to find out the difference between the conscious brain and the unconscious brain—because that will give you this unified field.

Sue But isn't this field worryingly close to magic, like a sort of psychic field...

John No

Sue ... or an extra force or...?

John Maybe field is the wrong metaphor, then, if it sounds that way; what I mean is this: it is a remarkable fact about consciousness, not only that there is a qualitative feel to any conscious state, but that you can only have it as part of a unified whole.

Right now, for example, I don't just hear the sound of my voice and feel the shirt on my neck, but I have both of those feelings as part of a single unified conscious experience that includes the sight of you and the sound of your voice, the view of the mountains and the palm trees outside the window, and so on; I have all of these as part of a single conscious field. That's why the stuff that Mike Gazzaniga and all those people did on split brains is so interesting to us, because it suggests that if you cut the corpus callosum you get two different conscious fields inside one skull, and I've asked Mike point blank, does he think that the research shows that. He's very cagey; he says: I haven't found an experimental way to show that, but certainly that's a possibility—that you might have two conscious fields that communicate, but don't coalesce.

Now maybe in normal life we really have two fields and they coalesce as long as nobody cuts our brain in half. So I don't mean there's something mysterious about this field; I don't think there's a sort of field of spiritual forces—like magnetism but more touchy-feely, or maybe less touchy-feely. That's not it at all. I'm just trying to give a verbal description of what is a defining characteristic of conscious states, namely that they hang together.

Sue **You mentioned there whether people with split brains have one consciousness or two; some people think that there's neither one, nor two, nor many; that in a sense there's none; that the whole idea of a unified consciousness is an illusion.**

John The marvellous thing about consciousness is that if you have the illusion that you're conscious then you are conscious. See, the normal appearance/reality distinction doesn't work in quite the same way for consciousness as it does for other phenomena. In other cases you have the appearance of something but there's a reality behind it which can be different from the appearance—it looks like there's a guy out there in the trees, but it's really just the play of the light and the shadow. But where the very existence of conscious states is concerned, you can't make that distinction; you can make distinctions within it— there may be some features of the field that you're misreporting or that you're not accurate about—but the very existence of this conscious experience, I can't be mistaken about.

Sue **I understand what you're getting at there. At least, I think what you're getting at is this: because what we mean by consciousness is how it feels to me now, then if I say, 'This is how it feels to me now,' that's it, no one can argue with me. And yet it seems that many people who haven't thought about consciousness very much just assume such things as that we have a full awareness of the visual world around us, or that we have a continuous consciousness; but when you look at some experiments or start to introspect very carefully, that impression begins to fall apart.**

John   I entirely agree with that. We have this unfortunate history where Descartes made it seem that the basis for all knowledge was the certainty we have of the character of our own conscious states—but we know from all kinds of experiments that there are all sorts of ways in which you can trick people, or in which people can just be mistaken in describing the character of their conscious states.

Sue   So you don't agree with Descartes then?

John   Absolutely not, no, no—there are a very few things about which I agree with Descartes, but certainly not this. I think we make all kinds of mistakes about conscious states. A lot of them are just because you're being inattentive or you misdescribe something; but there are deeper reasons—there's self deception: people are unwilling to admit they feel jealous; they are reluctant to admit that they're angry— 'WHO ME, ANGRY?' So I don't doubt that there are all sorts of ways that people can be mistaken about their conscious states; but the dimension of the mistake is different from the dimension of the mistake where you have features of the external world that you're misjudging. It is not the standard distinction between appearance and reality. There's a sense in which the appearance—if it is really *this* appearance and not something other—of the conscious state *is* the conscious state itself. You can be mistaken about the details of your present conscious state, but you cannot be mistaken about its very existence.

Sue   Do you think you have free will?

John   Well, I don't have a choice about that! We all think we have free will, and there's no way we can think away our own free will, because even if you try to think it away in a decision-making situation—if you just say, 'Well look, I'm a determinist so I just wait and see what happens'—that is itself intelligible to us only as an exercise of freedom.

   Immanuel Kant pointed this out to us a long time ago, that it's characteristic of conscious decision-making that you can't proceed except on the presupposition of free will; and that even if you try to deny it— if you say, 'Well, I don't believe in free will so I won't do anything'— that is itself only intelligible to you as an exercise of free will.

   However, you've got some interesting problems, because free will is not a characteristic of all consciousness—I don't have free will about seeing the table lamp if I look in that direction, but when I go to the restaurant and I look at the menu, I might decide 'Well, I'll have the spaghetti,' but I'm not forced to have the spaghetti; the other options are open to me; I could have done something else. So we can't think it away or pretend that we don't really have free will. The question that we don't know quite how to answer is, how could free will exist in a biological beast of the sort that we are? That is, if we have free will there must be something in the brain that is correlated with that free will, and what the hell is that supposed to look like? I have a lot to say about that, and we don't want to talk all night, but let me say a few things.

If you look closely at the experience of making up your mind and assessing reasons for action and deciding on one reason rather than another, there's a remarkable thing that happens: the considerations do not act on you like a set of forces that will produce a vector, like Newtonian mechanics. So suppose I had five reasons for voting for Clinton and three reasons for voting against him—I like his handling of the economy, I think he'll have a better foreign policy, he went to my old college in Oxford (he didn't in fact, but let's suppose he did), and so on. I have all these reasons, but I don't sit back and let them operate on me; I decide which one I'm going to act on. How's that possible, what's going on here?

I think we can make sense of that process only if we presuppose that this unified conscious field that I've been talking about is not just a bundle of disconnected perceptions of the sort that David Hume described. You have to presuppose rational agency; that there is some entity that is capable of decision-making, weighing of reasons, and acting. Traditionally that notion of a conscious, mindful rational agent has a name in philosophy; it's called the self—I hate this jargon, but anyway there we are. So in short, I think that you can't make sense of the experience of free action without postulating a self.

Sue You've got real problems there, though, haven't you? I mean, this self: what kind of a thing is it?

There are many scientists who would say you don't need that notion of self as a causal agent; that the real causal factors are all these interacting neurons which do many things, including creating a sense of self and a sense of free will, both of which are illusions.

John That's right, the whole thing might be an illusion; but let's figure out exactly what the illusion is. I've argued two things so far with you. One, that you can't make sense of rational decision-making and acting except under the presupposition of free will, the presupposition that there's a gap between the causes that operate on you and the actions that you perform. And two, that you can't make sense of your operation in the gap except on the presupposition that there is some *x*—I don't even have to call it a self—which is capable of thinking, deciding, choosing, and acting. That's all the self I need; it's not some mysterious mental entity; it's not a soul; you just have these logical constraints on the process of rational decision-making. OK, now let's suppose I'm right in describing all that, then that has to be going on in the brain somehow. Then you have two options.

Let's suppose the brain is just a total mechanical hunk of junk, like a car engine only wetter, and that it functions by absolutely straight-forward mechanical connections. Then what you would have is inde-terminacy at the psychological level—but it wouldn't make a damn bit of difference, because what was going on in the plumbing would be sufficient to determine everything you did; every move you ever made in your whole life would be determined entirely by causal processes—by complete determinacy at the neurobiological level.

That's one option. It's got a name; it's called epiphenomenalism—the mind doesn't really make any difference, it's just going along for the ride. That might turn out to be right—and if so, nature has played the biggest trick on us in history; that would be a bigger revolution in our thinking than Einstein, or Copernicus, or Newton, or Galileo, or Darwin—it would alter our whole conception of our relation with the universe.

But it doesn't seem to me that that's the way nature works; it would be miraculous if evolution created this incredibly complex, expensive apparatus, the conscious brain, if it made no difference at all.

The other possibility, which I think we can't rule out, is that the brain mechanisms create a system capable of rational agency that acts under the presupposition of freedom. That fact is itself mirrored in the underlying neurobiology, so the whole system moves forward through decision-making and voluntary action in a way that is con-strained by the conscious rational agent; and that conscious rational agency reaches right down to the bottom level, right down to the synaptic cleft. Now what the hell does that mean? I have no idea; I'm just telling you this is where the problem of free will comes out.

Sue Well, I think that problem is so dreadful that I wholeheartedly embrace your first alternative; I think nature has played this enormous joke on us, a joke well worth laughing at—here I find myself in this extraordinary universe with this illusion that I'm acting, when in fact it is just...

John   ... it's all mechanical, yeah.

Sue   Yeah. And when you say it's impossible to live without that sense of having free will, I dispute that. I've tried very hard, and to some extent succeeded, in living without that sense. And it does gradually go away.

John   I don't think you can live without it, because you can't decide what you're going to say next. Think of the difference between yourself acting and watching an old movie of yourself: when you know what's going to happen next on the screen is entirely fixed in advance. You don't, as you watch the movie, think, 'Well, I did a stupid thing then, let's hope this next time I won't do it'—because you know it's all laid down in advance.

Sue   I think that's an unfair example, because you're introducing a time lag. The whole point about this determinist sort of theory is that the decisions will be made; they will be made not by a conscious rational agent, but by all the underlying processes.

John   But those aren't decisions that make any difference; what you will have is a series of mechanical processes that determine events. It'll be exactly like an unconscious zombie operating with a series of clunky gears and wheels, a wind-up toy. And it might turn out that way—but the problem is, we've always been taught to believe it must turn out that way, and I want to say no, there's another possibility, and that is that indeterminacy at the psychological level is matched by an exactly isomorphic indeterminacy at the neurobiological level. Maybe that's going to turn out wrong, but it's a possibility that we have to consider.

Sue   Tell me about the zombie.

John   The zombie is really a philosopher's invention, to imagine a machine or a creature that behaves the same as a person who is conscious, but has no consciousness; and I think that makes sense; you can imagine such a thing; I can imagine that you really are a wind-up mechanism and that you're not conscious. It's a good thought experiment to imagine the difference between ourselves, who have both consciousness and coherent organized behaviour, and the zombie that appears to have the same organized behaviour but does not have any consciousness, has no feelings.

Sue   Obviously it's possible to *imagine* such a zombie, but are you saying that such a zombie could in fact in principle exist?

John   In principle, sure.

Sue So as far as you're concerned, then, there's something extra; you could have a mechanism that did all this stuff, but it wouldn't be really like us; it needs something extra, the conscious field or the rational agent or something like that, to make it be like us and have our kind of awareness. Is that what you're saying?

John That's exactly what I'm saying. I think evolution probably could not have produced such a thing, because evolution produced us. You can imagine evolution producing beings that moved around on wheels instead of on legs; but for all kinds of reasons it's unlikely that evolution would ever be able to produce that. Similarly, you can imagine evolution producing a well-organized zombie, but it's unlikely; we just get this much more efficient mechanism if we have consciousness. However, you could, in principle at least, design machinery that could behave as if it were intelligent—that is, could behave in the same way that human beings behave; we're nowhere near being able to do that, but in principle it's possible.

Sue I guess this relates to your thought experiment of the Chinese room. Would you mind summarizing the Chinese room for me?

John There used to be, I guess there still is, a view about the mind, which said that the brain is really a digital computer, and the mind is really a computer programme.

That had two consequences: one is that we would completely understand our minds if we figured out the programmes we were operating on; and two, we could artificially create minds just by designing the right programme. I offered a very simple refutation of that—so ludicrously simple I assumed that everybody must know it; but they didn't; it turned out that a lot of people were quite surprised by it—and here's how it goes.

I don't speak Chinese, in fact I'm hopeless with Chinese—can't tell Chinese writing from Japanese writing. Imagine I'm locked in a room, where I have a programme for manipulating Chinese symbols, and I get questions sent into the room in the form of Chinese symbols. I look up in the rule book what I'm supposed to do, and I give back answers in Chinese; so I take in Chinese input, and I produce a Chinese output; all the same I don't understand a word of Chinese. Furthermore, if I don't understand Chinese on the basis of carrying out the programme, then neither does any other digital computer on that basis, because no computer has anything that I don't have.

Now then, if you contrast my behaviour in Chinese with my behaviour in English, you find that in English my answers are as good as a native English-speaker, because that's what I am; if they ask me questions in Chinese my answers are as good as a native Chinese-speaker, because I'm going through the programme. On the outside it looks the same, but on the inside there's a tremendous difference—what is it? Well, why not just state the obvious fact: in Chinese I don't understand any of these words, I just carry out the steps in the programme. In English I have something more than the steps of the programme, I actually understand what the words mean.

This is so obvious; the computer doesn't have to know what anything means; it just works by manipulating symbols, zeroes, and ones; but the mind has something more than symbols; it has semantic content. This view I refuted is called 'strong artificial intelligence', just to have a label for it, and you can summarize what was wrong with it in four words: syntax is not semantics.

Sue I'm tempted to launch into all the famous arguments. I'm going to resist ... but will you just tell me what's happened in the intervening years?

John Well, I was amazed at the reaction that this argument provoked; there must be hundreds of published discussions of it. There are intellectual reasons having to do with the whole world view connected with the computational theory of the mind—it was a sort of reductionist attitude to consciousness and mental life, which went with behaviourism, functionalism, and great excitement about computers as the key to understanding human beings. And something else I discovered is that I was threatening a lot of research grants, and careers and money. We don't worry about this in philosophy because philosophers never get any money anyway, but an awful lot of people had big research grants based on the false premise that they were creating minds. So there was a continuing battle, which still goes on—and I think will go on until this generation of artificial intelligence people passes on and a new generation comes in.

Sue I said I'd resist the temptation to go into the arguments against...

John There are a number of arguments that keep cropping up over and over and over, and the one that became the favourite is, I think, actually one of weakest—but here's how it goes.

I'm there in the Chinese room but I'm not alone; I have the rule book, and a table and a desk, and paper, and boxes full of Chinese symbols; it isn't me that understands Chinese, it's the whole room, the whole system, that understands Chinese; I call that the systems reply.

I think that's kind of a desperate move, and I'll tell you exactly why—I'll answer the question, why don't I understand Chinese? The answer is obvious: because I have no way to get from the syntax to the semantics; I have no way to get from the symbols to their meaning. But if I don't have any way of getting from the symbols to the meaning, neither does the room.

Just imagine that I put all the room inside me; imagine that I memorize the rule book and all the symbols—it's science fiction anyway, it's a fantasy—but suppose I do memorize all that; now get rid of the room, and I work outdoors in an open field and do all the calculations in my head; then there isn't anything in the system that isn't in me— and I still don't understand Chinese.

I think the fact that so many people fastened on to that is a sign of desperation. What they should have said to me, on their own terms, is, 'Of course you understand Chinese, because you passed the test for understanding Chinese and you had the programme.' That's too ridiculous; very few people have the nerve to look me in the face and say 'You understand Chinese,' or, 'You would understand Chinese in the Chinese room.'

Sue D'you know, I am so delighted by that answer, because what you call that last brave answer is the only one that I've ever thought of: that *if* the room operates that way in this thought experiment, then you *must* understand Chinese; because that's what we mean by understanding Chinese. You may say it's ridiculous and a last resort, but I'm glad that we at least agree that it is a reasonable response.

John Oh, I don't think it's a reasonable response; I think it's crazy, but I think it's courageous. The real problem is that the computer has no way of getting from the syntax to the semantics; another way of saying the same thing is that simulation is not duplication. We can simulate anything—the digestive processes in your stomach or the flow of money in the British economy; but if we do a perfect computer simulation of digestion nobody thinks, 'Well, let's go get some fish and chips and stuff them into the computer and see if it'll digest'; it won't, it's just a model or a picture. The computer simulation of the mind stands to the real mind the way the computer simulation of digestion stands to real digestion; it's just a simulation, it's not the real thing.

Sue Now I just want to go back to evolution. It seems to me that if you believe zombies are possible, then consciousness is something extra; and therefore if we think about the past evolution of human beings, we have to say that selection pressures would have had to favour consciousness; in other words, consciousness must have a function.

John That's exactly what I'm saying. Consciousness has an enormous number of functions. Take our present mode of existence: we take in all of this incredible amount of information, and we organize it in a conscious field; we slough off the information that we think is irrelevant; we then collate the information and organize it to make decisions. This gives us much greater power, flexibility, ability to process information, than you would get if you just had simple unconscious mechanisms; and that unifying feature that I talked about earlier is crucial here, because I now have visual information, tactile information, auditory information, and memory information, all coordinated in a single conscious field; that's a very efficient mechanism.

Sue And you think that natural selection has acted upon that, to improve it as it's gone along?

John I think it's no accident that conscious beasts tend to do rather well in the struggle.

Sue So what's a conscious beast?

John Well, we are one.

Sue I accept that much; but what other beasts, what kind of beasts?

John Well, we don't know; since we don't know how the brain does it, we'll have to wait for the experts to tell us whether or not termites are conscious; my guess is they probably are.

Sue But how could any expert ever tell us?

John I'll tell you exactly how. Suppose we discover that there are very specific brain processes that cause consciousness—so that, for example, in brain-damaged patients we can re-introduce consciousness by artificially producing certain kinds of brain mechanism. To give these mechanisms a label, let's just call them XYZ: it's XYZ that causes consciousness. Now we go down the phylogenetic scale and we discover, no question, that dogs and cats and primates all have XYZ; but when we get down very low we discover that termites have XYZ but snails don't. And furthermore, let's suppose we have another explanation of the snail's behaviour, then we'd have to say, 'Well, OK, snails are not conscious and termites are.'

Sue   And would you say then we'd have solved the problem of consciousness?

John   If we got that far, we would have; it would be a prodigious intellectual achievement, yes.

Petra Stoerig

It's obvious that other animals experience very much like we do

Petra Stoerig studied philosophy in Munich and then gained her PhD for work on the mind-body problem there in 1982. That led to work on neurophilosophy and medical psychology, as well as research on the phenomenon of blind-sight. Her research interests include the neuronal basis of consciousness, neurophilosophy and conscious vision; she loves opera and animals and has a special interest in ethics in science and medicine. She has worked in Oxford, Montreal, and several universities in Germany, and holds the Chair of Experimental Biological Psychology at Heinrich-Heine-University Düsseldorf.

Sue I want you to tell me what you think the problem is; why conscious-ness is so interesting and difficult.

Petra Why it is interesting? That is an easy question: of course it's interesting because it's the basis of all our joys and pleasures and pains and tragedies and so on. Without consciousness there wouldn't be any experience. From my point of view, consciousness is the pre-requisite of experience, and experience is what makes us happy or sad or enjoy the sunset or a glass of wine, or a wonderful Belgian meal...

Sue Why is that so difficult for scientists?

Petra It's very difficult to explain how it comes about, and why we have it. I think we have it because it serves a survival function and it's a very big motivator—after all, human beings have been known to climb Mount Everest, or travel to Sao Paulo to hear an opera because it's not performed that way anywhere else in the world, and it's unimaginable that they would do that without having qualia, without having experiences. That's exactly what the magic of life is about—so in that sense of course it's the most interesting thing that there is around. Now, why is it difficult—because obviously the question is, why does consciousness come about, what is it good for, how is it made?

Sue And do you have any idea about that? Because I don't think that it's good for anything, but perhaps you think it is.

Petra I think it is good for survival; it's the big motivator; it's the thing that lends magic, and is conducive to life.

Sue You mentioned qualia, so I assume that when you talk about consciousness you're talking about subjective experience; can you tell me more about what you mean by the word qualia?

Petra No, but I can tell you more what I mean about experience, because there is no experience without qualia.

Sue But I don't know what you mean by qualia.

Petra But you know what I mean by experience?

Sue Yes.

Petra See, that's what I mean.

Sue That's all?

Petra For me consciousness is experiencing.

Sue So you don't mean something like the philosopher's idea of qualia as ineffable, irreducible qualities of...

Petra But that's part of experience. If I see your green hair, that's green, and we can agree about it even if there are subtle differences in our perceptions. It is only if you can't see colours, or taste a wonderful wine, that it is close to impossible to make you understand what it is like. In that sense it is ineffable, and in that sense it's also irreducible. Only oneself gets directly informed about one's experiences, and that makes it a philosophical issue. We can't imagine what an organism with senses other than ours 'feels'.

Sue You said that it has an evolutionary function, but it seems to me very peculiar to imagine that this ineffable greenness of my hair could have

any effects upon anything. It's ineffable, you can't say anything about it; it's just your private experience; in what way does that have a function?

Petra How can you possibly say that I can't say anything about it, and when I'm obviously...

Sue ... 'ineffable' means you can't.

Petra ... but I can talk about it; I can describe its shades and its differences; I can try and describe why I think I like it, or why I think it doesn't suit you, or whatever. I can say how it goes with your shirt; I can say all kinds of things about it—of course I can talk about it, and I can talk about it to someone else who sees the same thing.

Sue Can you tell me how you set about investigating this in your own work—consciousness I mean ?

Petra Yes. I'm studying a particular case called blindsight, in which the phenomenal feel to things is lost due to a brain lesion which causes cortical blindness. The patients I most commonly work with all suffer a lesion to the primary visual cortex or its afferents. This causes cortical blindness in the entire visual field, or in the area where the functional or structural loss is. And because this lesion is at the back of the brain and quite distant, really, from the eyes; and because we have very many parallel pathways leading from the eye into the brain, we have a situation in which the patient tells you that he or she doesn't see anything any more in that part of the field.

Nevertheless, there is a lot of visual information coming into the system. You can find out that there is actually this information being processed and that patients are able to respond to it—and respond differentially to it. I've been studying this for a long time because I'm very interested in the question of what consciousness is good for; I have an example here in which I can study that: what really are the things that these patients can't do because they lack the subjective qualities, like greenness and so on, at least in part of the visual field.

Sue So these people all have a part of the visual field where they say they can't see anything, they have no qualia; but they can still do certain things based on visual information. What sort of things can they do?

Petra They can detect whether a target is there or not, they can localize this target, compare it with another...

Sue It seems like magic, doesn't it—that they could say, 'I can't see it,' and yet...

Patients with blindsight have a scotoma (an area of their visual field) in which they claim to see nothing. Yet when stimuli are shown there they can often guess accurately such details as whether a line is vertical or horizontal. Is this vision without consciousness?

Petra ... and yet they can even make a lot of discriminations. They can do discrimination of colour and size and orientation and motion—tell whether a motion is there or not—and so on. So they can do a lot of things, really; and with time they get better and better and can do more and more things. I'm not sure if this is really the road to understanding the role of phenomenal vision that I originally thought it might be, because the people are getting better and better; and maybe eventually—there's some evidence for that—they'll start seeing again; and then of course I won't know what the things are that they can't do. So that's a dilemma, but of course on the other hand I'm very happy if they start seeing again.

Sue Of course. But if, for example, you put a red flower in front of them, and they say they can't see it, because it's in the blind part of the field; and you then say, 'OK, guess: is it red or yellow?'; and they guess, and they're right more than chance; then you keep on doing this and they get better and better; do they then have any experience? Do they start to talk about what's happening in a different way; is there any hint that they're beginning to get colour experiences?

Petra I do train the patients, so this is different from other people studying blindsight. I think that I really train them: I have a lot of patients who can't do it in the beginning, who really perform at chance level, as you would expect from what they say. But some are very fast at learning, and some take longer in acquiring this capacity: some can really do it at almost the first trial.

I had one who underwent some testing of red/blue discrimination, and he was given feedback—that is part of the training, of course: upon every response he gives he learns whether it is correct or not. And he was looking at the screen, fixating as he should, and reading the feedback and going on tapping his keys, and saying, 'I can do that, I can do that—look, I can do that.' That was one of the fastest I've seen so far; he was a young man, so age may well play a role.

In the longest case it took almost two years before he was above chance; and then it developed normally; and once it has developed and you go on doing it—and I've been going on for a very long time with people—they may start to say that they have a feeling that there is something there. This feeling is not necessarily always correct— they may be wrong about that, and it's not very reliable in the beginning; but nevertheless they get better and better, not only in performance but also with respect to feeling safe, with respect to navigation. I have patients who ride their bike for 20 km to come to the lab for their weekly sessions, and so on and so forth—so it makes a big difference to their life.

Sue I think Dennett has called this 'super-blindsight'. He argues that if they were trained, and they could do it perfectly, they then should have the same visual experience as someone with normal vision. In other words, he thinks you can't separate the experiences from the abilities. Is that what you would expect?

Petra Well, I think I'm forcing the system to respond to information in the blind field, which still has a neural representation after all; and this really does something to the system. You force it and you force it to pay attention to something that obviously is subliminal at first; and that's why patients get better. Maybe other parts of the system get recruited into something that's been lost due to the lesion, and I think there's evidence that this may eventually lead to at least partial recovery of conscious vision. We don't yet know what parts of the brain really will do that, but it seems that this means it can't be, or very likely it's not, just a structural thing; it must be some kind of functional, or computational skill that the brain reacquires.

Sue Do you think of these patients as partial zombies?

Petra No.

Sue So they're not just missing qualia in the traditional sense of a zombie?

Petra Of course they are missing the phenomenal representation of the visual sense; but, after all, a zombie is someone who behaves exactly like you and me, and looks exactly like you and me, but has no inner experience whatsoever; and I think that kind of thing is biologically unviable, so that's why I'm a bit opposed to zombies.

In fact I hate zombies because there's so much paper wasted on a thought experiment. I think they are logically possible and they may be interesting in that respect for philosophers—well, obviously they are! But as a biologist I think it's a waste of all the trees that go into this paper, because it's not biologically possible; there is not a single being that we know of that's able to behave like you and me but with nothing inside.

Sue What got you interested in consciousness in the first place?

Petra I think it's the most important questions, really, for a human being. That is to understand these three things about consciousness, of which I've already mentioned two, namely: how is it made (which is not as important for everyone, but more for neuroscientists maybe); what is it good for (which I would like to understand); and the third one that I would like to understand is, who has it? I'm very interested in this comparative question of how can you find out which are the living beings who have it. We're pretty certain for everyone who's close to us, but we have just no idea how to test it in an organism that's very different.

Sue Have you always been interested in these questions?

Petra Well, probably not as a child. I think when I was a child I had this idea that I would solve the problem of cancer, and then eventually I had the idea that I would be better to solve the problem of consciousness. I'm not sure it was the right decision!

Sue And I'm not sure which is more difficult! Do you think that having tried to solve this problem, instead of the cancer problem, has changed you in any way; has studying consciousness changed your own life, your own consciousness?

Petra Well, it's changed my life, because it's eaten up a large part of it—in that sense, yes; but I assume that you're more aiming towards the question of whether I've become more conscious—and I actually don't think that's the case. I don't think it's changed my experiencing of the world. It's changed my view of people and of their divergent opinions, and I've learned much more about things like that, but I don't think it's really changed my experience.

Sue You've talked about evolution and other animals; has it changed your approach to other creatures?

Petra No, because I think it's so obvious that they experience very much like we do. Of course they don't experience the same way; consciousness consists of different aspects, and some of them they have like we have, some of them they have in a different way, some they have that we don't have. So of course it's not the same, but the basic principle that they do experience what they do, and what happens around them—I think that's definitely there, and I mean, that's part of why I love them.

Sue And what do you think happens when you die?

Petra Ah, there she comes with her next question! Finally you come out of the closet!

Sue I'm just asking because when Stuart Hameroff was talking about microtubules he claimed that quantum coherence carries on after the body's dead, and I thought readers would like to know what other people think.

Petra I don't agree with Stuart, but that's a very difficult question, and I really simply can't say anything other than that I don't know.

Sue Fair enough, but it's also very relevant to what we understand about the self.

Petra Yes, and at the moment I have this fancy about self-consciousness. This topic is really important for me. I've shunned it for a long time because it's always treated as something different from consciousness—and consciousness is already two things. It's being in a conscious state as opposed to being in an unconscious state, as maybe in coma; and it's about conscious content—what are you conscious of, and what are you not conscious of, at any given moment in time or in principle. Those are the two things that I've been dealing with so far, and I always thought, well, maybe self-consciousness is something different, because it's so often treated as if it is.

Then I started thinking a bit about the role of the self in biology, and the role of self in consciousness; and I think we go about testing it the wrong way; that's why I care about it.

I think many of us have this idea that self-consciousness is something that is reserved to the human species. Whenever there is a demonstration of it in non-human animals people use that mainly to redefine what they mean by self-consciousness: they say, 'Now, this has been very nicely demonstrated—that this or that animal recognizes it/him/herself in a mirror; but of course that is a far cry from being able to reflect about your own mental states and your emotions and your perceptions, and so on and so forth.' So with every new finding there's a sort of adjustment to what self-awareness or self-consciousness then is; I think they would go to almost any length just to reserve this to themselves.

Another question is, why on earth do they do that? That's really something I'm very interested in, because the self/non-self distinction is the most basic in biology; it's really the first thing you have to have if you want to not digest yourself; or if you want to tell that something is different from yourself, you need that.

Sue **You mean even a single-celled organism needs to have a membrane, and distinguish between inside and outside?**

Petra They have that. They have a very complex organization. Within this one cell they are performing everything that is necessary for survival: they have metabolism, they can find their conspecifics for having orgies—it's unbelievable, these things all being there in a single cell. Right?

Now, if you have a complex organism all these functions get segregated into sub-parts of this more complex organism; but that doesn't tell us anything, really, about whether there are functions like thought or consciousness in the single cell organism. There might be, we just have no way to find out. There might be.

Sue **So you've looked at all the evidence on mirror-self-recognition and other techniques; can you think of a way to make progress against all these people who keep moving the goalposts?**

Petra Yes, I think there are two ways to go, and one is to simply focus on the fact that there are very many aspects to self-recognition and self-awareness, and that if you want to test them in different types of species you had better adjust your question to the species.

So if a species has episodic memory, it's rather strong evidence for it having a self, because without the self you don't have any episodic memory. And there are many, many things like that.

If you look at how many species are able to move ever so much more elegantly than we do, if I may say so, then it's very likely that they have a good sense of proprioception and a kinaesthetic sense; and I think that is a basic form of self-awareness of which they may have more than we do.

I think there are plenty of other instances where they actually have more; and that not only makes me think that this is a kind of puzzle that has different aspects in different species (and also in different individuals, of course), but also makes me wonder why are we so jealous about this thing in particular?

I think it can't be because we're so superbly self-aware, because I think our self-awareness is utterly poor compared to our knowledge about the world; that if anything it would be protection of fragility rather than protection of something you really have a treasure of. I mean, there is no species around that knows as much about the world as we do—that's why we rule it; we manipulate it in every which way we can. But about ourselves, we really know very, very little directly—and this is about direct knowledge after all, self-awareness—we don't know our motives and we constantly tell stories to ourselves. This self knowledge is just so tiny and fragile, and it's very well possible that other organisms have more of that.

Sue Gosh, would you make a guess which ones those might be?

Petra I've been thinking a lot recently about what the reasons might be for this strange jealousy in so many humans about self-consciousness, and I think it might have something to do with language, but it would be really premature to talk about that.

Sue I just wondered—the most obvious species that come to mind would be dolphins or whales, or maybe elephants; but I wondered whether you had in mind others as well.

Petra I think that I would have to have a better idea of what prevents us from knowing more about ourselves. If, for instance, it was language, then it would be more likely to be animals that don't have indications of language, that are much more honest to themselves.

Sue But most people would say that language gives us more self-awareness and you're saying that it might be what gets in the way; what prevents us from seeing the truth?

Petra Yes, it might be what prevents us from seeing what really makes us move.

Francisco Varela

We're naive about consciousness, like people before Galileo looking at the sky

Born in Chile, Francisco (1946–2001) studied biology before moving to the USA for a PhD on insect vision at Harvard, subsequently working in France, Germany, Chile, and the United States. He once said that he pursued one question all his life. Why do emergent selves or virtual identities pop up all over the place? He was best known for his work on three topics: autopoiesis, or self-organization in living things, the enactive view of the nervous system and cognition, and the immune system. His many years of Buddhist meditation influenced his work on consciousness, and he was uniquely both a phenomenologist and a working neuroscientist, coining the term neurophenomenology. Until his death he was Director of Research at CNRS (Centre National de la Recherche Scientifique) at the laboratory of Cognitive Neurosciences and Brain Imaging in Paris. He has written and edited books on ethics, consciousness, and phenomenology and is co-author of *The Embodied Mind* (1992).

Sue   What for you is the problem of consciousness? What makes it so special and so different from other problems?

Francisco   Well, maybe it's not so different. Maybe it shouldn't even be called a problem—it is a major fact. I mean if I look at nature, just very naively speaking, there are two things that stand out: there is the world, and there is me. Simple, right?

If we are scientists, and by scientists I mean we want to understand the natural world, then the me-ness part—the us-part—the conscious part, is going to be exactly one half of the picture we need to account for.

Now, there are all kinds of reasons why this particular fact of the world has been left out in the history of science, but it's not a very kosher affair. In many ways it's political, and even in the twentieth century it has come back and been thrown out twice. In 1905 or 1910, with the rise of phenomenology and introspectionism in Germany, it was the thing of the day. Then by the 1930s, or certainly after the war, it was out. Now it's back in with a little neuroscientific edge, and maybe it's going to be out again in the next ten years.

Sue Why do you think it was thrown out early in the last century? Was it just too difficult for people to tackle, or was it some other reason altogether?

Francisco Look, I don't think that there is any simple explanation. I don't think it's particularly more difficult or less difficult than other topics. I would say that primarily it's been sociology of science factors that play into this. For example, you have to remember that the war had just happened. Everything that had to do with the study of consciousness, such as the phenomenology of Husserl and Wundt, was identified with Germany—and through Heidegger with the Nazis. So after the war, because Europe was shattered, research took off in the United States, and was rapidly dominated by people who had then become the famous behaviourists of the day.

But in the United States there was also William James. Interestingly, everybody today cites William James like the Russians used to cite Marx and Engels in every speech. Yet William James is an interesting guy. His *Principles of Psychology* is very much in the modern spirit of scientific investigation. And in his later work, in *The Varieties of Religious Experience*, and in *Pragmatism*, he is really far out—much more far out than anybody makes him. In his later work consciousness is the very substance of the universe. So the basic fact of what it is that exists is grounded in consciousness. According to him, consciousness is not reducible to explanations that come out of biology and neuroscience.

So you see I'm quite puzzled. I don't think that there is a simple explanation but we have to live with the fact that there is heavy resistance. Now one kind of resistance, that is also part of the explanation, is that to study consciousness you need the data that goes with it—first person data.

Sue You immediately hit trouble there don't you? Normally when we talk about data we mean data from the outside—data that are publicly available—but you are talking about data from the inside. Haven't you got a problem here?

Francisco Of course that's the usual take; that the data from the outside is reliable and the data from the inside is subjective and fuzzy. Well, is it true, or is it just something that goes with a reluctance to really study consciousness?

You see, if you think about so-called objective data in physics or biology, nothing is ever going to be observed unless you have somebody that reports on it. So you inevitably have a first person component. That's the first element. Second, the fact that you're reporting data and then it becomes so-called 'objective', is because that report is inter-subjectively validated by other people. This means they can use the same protocol, and go and look under the same conditions, and that's the stuff of science.

Now I ask you this—when you have reports of data that are accessible through first person methods, and you put it out for inter-subjective validation, why shouldn't those accounts also be equally valid—and become part of the common knowledge? So the distinction between objective/subjective is merely what? A change in the kinds of tools you use to observe.

Sue But is it? Let's take colour, for example. You're wearing a beautifully bright yellow shirt. Now I have a very strong impression that my subjective conscious experience of that yellow shirt is something completely private to me. All I can do is tell you I see yellow, but I can't in any way convey the 'what it's like for me' of that yellow. Isn't that a problem? Isn't that something about which we need to know if we're going to talk about consciousness—but we can't get at by a third person approach?

Francisco Well, that's precisely the point. Should we distinguish the quality of privateness from the quality of access? It is true that only you can tell me what is your experience of my shirt, but that doesn't mean that it is private. Why? Because you can report on it, and that report can be inter-subjectively validated. So that if I say, 'No, no, it's not yellow, its red,' then we can either disconfirm or confirm it, like you do in other sciences.

Sue But isn't there something basically different about this? I agree I can tell you I see yellow, but that's not what I'm getting at. I have this feeling that for me yellowness is something that I can't communicate;

the word yellow doesn't do justice to it. Or take something deeper than that—say I really feel emotionally moved by something, or have a feeling that is important or profound for me. Then any words that I give to it to tell you about it, they don't do it justice.

Francisco That's fair enough. You raise two points that are at the core of the difficulty the scientific community has with consciousness. One is to me less profound than the other.

The first one is what I would call the methodology problem. OK, when you say it's not enough just to say 'yellow', of course it's not enough just to say yellow. To do good accounts of what you experience is not a trivial affair. In fact, if you do that with normal subjects—if you bring them to the lab and ask them about emotions, you ask 'What are you experiencing?'—most people go blank. It is not given to man to be experts of their own experience; the fact of having an experience is not a qualification to be an expert reporter on it, just as much as walking in the garden doesn't make you a gardener, or a botanist. You need to have a very substantial amount of training.

This is, to me, one of the core resistances in the West. We have one kind of method that has come from the scientific tradition, but I really think that we have to look and understand the accumulated empirical and observed knowledge in other traditions. I'm particularly interested in the Buddhist tradition where sophisticated methods of training subjects gives you the possibility of actually reporting on your emotional life, for example, in extremely precise, sophisticated, and inter-validatable terms.

So that's the first point. One of the reasons that it seems so flat just saying yellow is because the richness of the description is just not there. And in order to have access to that we need to introduce new first person methodologies way beyond those we have at the moment, and that means a sociological revolution in science. Among other things you have to train young scientists to become proficient in the techniques, you need a complete change in the curriculum design and so on. You know, I think we're extremely naive. It's like people before Galileo looking at the sky and thinking that they were doing astronomy.

Sue And the second point?

Francisco Once you have the method you have to explain the phenomenon as such. It is the 'what is it like it to be'. So the question is, why is it that consciousness feels so personal, so intimate, so central to who we are, and of course, that's why it's interesting. The study of consciousness is a kind of singularity in science, because you're studying precisely the most cherished quality of what it is to be alive. So the second bit has to do with how to account for that intimacy. Now that's a different problem and I think that progress in doing that has to come by understanding how the brain works; how it can differentiate colours and forms, and have motor programming, and have different kinds of emotions. All that machinery is not just like in a computer, where it has to produce some result. It is a device that evolved over a long period of history, both phylogenetic and ontogenetic. It only makes sense in the context of being active in the world, and that embodiment is precisely what we experience.

We experience ourself intimately because we're embodied. Therefore, the state of consciousness as a pure mechanism won't do; the mechanism is a condition of possibility to give rise to something that feels like somebody because it is embodied.

For example, I'm touching a piece of the bottle here. That bottle feels bottle-like, that is solid and immoveable and obstructive, because when I touch it that's the quality it has. In other words, the physics of the world is such that solidity is what allows you to do certain things and not others.

Sue   **But still the hard problem seems to be lurking in there. I mean, we can describe you holding that piece of bottle in one way in terms of neural potentials going up your arm and into the brain and so on, or in another way in terms of how it feels from the inside. So how is it that 'bottle feelingness' comes about from this embodied relationship between some neurons in a brain and the bottle out there?**

Francisco   This is the reason I call it neurophenomenology. The neuro-part gives you a fundamental insight into how the brain works, but it won't give you the -pheno part. The -pheno part requires both putting it into this embodiment and having the first personal access to report what it is like. And it is the combination of these two that will do it. In other words, my claim is that you cannot do without one or the other. The whole point is to get used to thinking and doing science in a different way, by combining these two things.

The reason I use solidity of an object is because the way we handle objects is so well studied in neuroscience, but at the same time the idea of embodied action is also a very rich theme in phenomenology. So, when you combine the two, all of a sudden it's like looking at things from two perspectives; it becomes 3-D. There is no longer this contradiction that the hard problem claims. The hard problem is going to be hard only if you stay hemi-blind.

Sue What about that computer you mentioned? Suppose you had an embodied computer, a robot that had hands and could pick up the bottle just as you did. Would it necessarily, in your view, have subjectivity?

Francisco It would not have subjectivity that is akin to ours because we have such a long evolutionary history, but yes, it would be on its way to having it. It might have a kind of primary consciousness like that in a cockroach or in a dog. So I don't think that consciousness needs any kind of extra ingredient.

Sue Then can I ask you the classic zombie question? From everything you've said, do you think there could be a creature that could do everything that you do, behave exactly like you do, and say the kind of things you do, but have no experience inside?

Francisco Susan, I've always had the hardest problem with the zombie argument, because it seems to me that it's the typical kind of argumentation that happens in the Anglo–American philosophy of mind tradition, which is really not my tradition. I just don't grock it; I don't get it. Of course you can imagine that such a thing would be possible but it seems so absurd to imagine it. I say it's just a problem that you create by inventing problematic situations. So what?

Sue I'm surprised, because I thought you would say that the answer is obviously 'No'. Doesn't that follow from everything you've said so far about embodiment and behaviour?

Francisco Well, you might want to say that, but the problem is that the zombie people assume this thing does not and will not have conscious experience; then they're stuck with this imaginary situation that doesn't really work for me. From my standpoint, it's an open empirical question.

Sue But is it? I thought an empirical question is always one where you can get the answer. Yet if you had a well developed robot, and it was going around saying, 'I'm conscious, I can feel the experience of that yellow shirt,' it would be easy for someone to say, 'But we don't know that it's really conscious.'

Francisco No, you see that's where the weight of the tradition shows up in the way you phrase the question. Because if we have an intrinsic problem with knowing when somebody else is conscious then we couldn't live.

My counter argument would be that being human, and being alive, is knowing profoundly that those around me are conscious. And the idea that I have to convince myself that you are conscious—and you're not a zombie—is just ununderstandable. It's just complete nonsense, because I am built from the ground up by this impossibility of having a consciousness which is identified with Francisco without having Susans and Jims and Joes around the world.

In fact we have very strong empirical evidence for that in the way babies develop; that the awareness of one's body when you're a tiny little baby is fundamentally built on the understanding of what it is to have a body for the other person. And notice what is happening now with higher primates; the more people work with these creatures, and the more empathy has a chance to develop by living together, the more you have people like Sue Savage-Rumbaugh working with the Bonobos. She has absolutely no doubts that their experience exists.

So if we have robots that eventually grow around us, that's what is going to happen. Like in good science fiction we're going to be able to tell when a robot is of the conscious kind and when it is built to be a stupid little slave cleaning rugs. So that argument doesn't carry any force for me.

Sue   I do enjoy the way you reject some of these arguments that have been so deeply embedded in the tradition. As you rightly say, for me they are a problem but you're making me realize that we don't have to think of them that way.

But let's turn from the way it has been thought about, to the way you want it to be thought about. You've talked about training scientists in a different way, and about learning disciplined use of the first person perspective. How do you suggest this can be done? Is it already being done, and how should it be done?

Francisco   It has been done in the following sense: that you do have on this planet a small percentage of people who are highly trained subjects, in other words who have spent years learning how to describe what experience actually is—for experience is not something that is given to you immediately, it has to be unfolded. Like anything else in the world which is complex, just the first glance won't do. The problem is that most of these people are not scientists. So the only way you can get them is to bring them into the lab, and do experiments with them as collaborators. So, for example, I take people who are trained for 20 or 25 years in the Buddhist tradition of meditation. You can ask those people questions that you cannot ask a normal person, or you can ask them to do tasks that are normally impossible, such as to keep a steady attention over say a 25 or 30 minute span.

You know, it has been reported in the literature that the US college population only has an attention span of about two and a half minutes at the most. So the metaphor is simple: if you have a flickering light, like a candle in the wind, you'll only be able to observe while the light lasts. Now if you have somebody who has stability of attention then it's like a light bulb that can be sustained for 20 minutes. So we're going to see different things. That's the point.

Now it's not going to be my generation. It's going to be the young people who get enthusiastic about this paradigm of neurophenomenology and realize that they themselves have to acquire that learning. So then in the next generation these competences can be combined.

Sue But there are such people now aren't there? I mean, I wouldn't put myself as a fantastic example but I've been meditating for 20 years and, even if I can't do it perfectly, I can usually sit and maintain attention for 30 minutes. I know there are some other scientists who are trained in Buddhist traditions but as far as I can see nothing very earth shattering has come out of that yet. What is it you want to do with such a person? I mean, if you had me in the lab, and if my claim were true—that I can sit there and concentrate on something for 30 minutes—what would you want to do with me?

Francisco Right, I agree with you, and like you have been a meditator for 20 or 25 years now, but I didn't put it that way because it is a little better to start where you're not qualifying yourself as it were.

What would I like to do? The first step is a very simple one. Let's go back to doing simple experiments like, say, perceiving a face. Right now, studies of perception, memory, attention, and so on, are all based on having a population of subjects, and averaging out the results. What I'm trying to do is to take a highly trained subject and the same basic task, but now you take presentation after presentation, and after each presentation you ask them to give you a specific report of what happened in that individual trial. So you can have exactly the same paradigms that we have been studying for years, but this time you get the entire gamut of mental conditions and mental states, or else a very homogeneous and highly stable set of mental states.

First of all I want to see whether you get the same neural correlates or not. And I can tell you that you don't; that when you take such reports and you separate out the different trials, the correlates for different mental conditions are totally different.

Sue What do you mean by the neural correlates there? Are you talking about brain scanning or EEG or...

Francisco I left it intentionally vague. I work with EEG and MEG correlates because I'm interested in things that are relatively fast but you could do the same thing with PET or with MRI.

Sue So what have you found out about the difference between ordinary college students and trained subjects in terms of measurable differences in their brain?

Francisco Well, let me give you just a quick example. In our lab we have studied stereoscopic fusion, in which subjects have to see a 3-D image. Now typically this is a relatively long task, and subjects take their time to build their strategy, but when you look at the trial-by-trial data, you discover that it is extremely variable. We recently had a highly trained subject in the lab, and he reported going into this state of having absolutely no thoughts. Nevertheless, he could do the task quite precisely and press a button when the fusion came out. What we found was that his brain activity was absolutely clean. The brain sites and the frequency bands that are active are reduced to just one; essentially the one related to the motor response. So it's very interesting to see a brain correlate of somebody who is not having any thoughts or distractions. So there you are—that's a question that you couldn't ask otherwise. 'What happens in the brain when there is no thought; when you just have primary consciousness, and not reflected consciousness implied in thought?'

Sue It suggests to me that each of us has this extraordinary instrument for doing things—our brain—and that most of the time, it's just flooded with nonsense and not being used at all effectively. Are you saying that when it feels from the inside as though the mind has calmed down, and thoughts have slowed almost to stopping point, that that's visible as reduced brain activity?

Francisco I would bet my hand that we are going to see those differences. I mean you do understand the point about training because you have been involved in it, but this is a real blind spot when trying to talk to your basic scientist who has never heard of the idea. It is difficult to understand that there is such a thing as training in having access to your experience. The concept itself is very foreign.

Sue Do you really find resistance from scientists when you try to do this?

Francisco Resistance is not quite the word; it is more like puzzlement. Of course there is the fraction that are downright hostile and thinks that this is just nonsense, but I would say a good deal of them just look at you and say 'Aha, interesting, mmm'. They're not against it but they just don't get it. So I think it is upon us, those who strongly believe that this is the direction to go, to start making some progress.

Sue I think I'd want to go even further than you in some directions! For example, I've been very interested in the research on change blindness, which clearly shows that every time we move our eyes, or blink, the visual world is just thrown away. Now that's very strange and it seems to conflict with ordinary everyday experience. So I have sat for long hours in meditation watching and asking the question 'Is it like that? Is it all thrown away?' and it seems to me that it is. In other words, the experience changes to become more like what you would expect from the change blindness findings.

I am able to sit and look at the world in such a way that things just pop up and disappear—another thing pops up here and now—and another. In a very strange sense the stability is completely gone, and yet it's not particularly disorientating. Now this, to me, would be a way of combining first person and third person work. But from what you're saying I guess that would get even more resistance than the kind of experiments you're talking about.

Francisco Yes. That's why it's a lot better not to use yourself as the subject. You see I think we have to go very slowly in this path and do simple things like stereoscopic fusion, that we can link back to more traditional studies, before we study the more interesting things like no thought.

Sue Oh dear—now I find myself wanting to go further than you which quite surprises me. I mean, I don't expect to be able to get up at a neuroscience conference and tell people about this. Nevertheless, it seems to me to be essentially part of my work as a scientist trying to understand consciousness, that I work on myself as well. Otherwise how am I going to make sense of this?

Francisco That's absolutely true. I don't think that I would have thought of any of this had I not been involved in this kind of work myself, and I use it as a primer to ask the questions I ask, and to choose the subjects I do. But, you see, we're talking about a sociological phenomenon. We have to respect the rules and move with the community, so that we're not treated in a marginal way.

So, personally, the drama or the joy, or both, of my life is that I have one foot in one side, and one foot in the other, and I refuse to marginalize myself. And I refuse to shut up the side of me that knows that this examination from the first person is possible, and essential.

Sue **You're quite an unusual person really because you combine science, and a scientific background, with an interest in phenomenology, and a French background. So what has studying all this done for your life?**

Francisco Well Susan, to me it's almost the other way round. I started my inner work—if you want to call it that—for the same reasons that anybody else does in the Buddhist tradition, which is confusion, pain, and disarray. On top of that I had a bit of a civil war on my back—the kind of situation where you say, 'Well I don't think I understand very much of what is going on.' Then it took me about ten years to realize that behind this practice of meditation to quieten your mind, there was a Buddhist theory of mind. This was fascinating; like a treasure trove of humanity that these people have kept alive, and brilliantly expressed and analysed. That's the point when the first person tradition affected my professional life, by making me think that what we were doing wasn't quite right. It led very explicitly to this notion of embodiment that I expressed in my book *The Embodied Mind*. I did it to move cognitive science away from the idea of information processing, into this embodied or enactive perspective, which is now picking up quite nicely. This already led me to change my own way of doing science, and now this neurophenomenology formulation is a second step in that direction.

So in my life it's as if I started out with these two things being completely apart from one another, and by now it's hard for me to say who is who, and I'm more unified. Now this poses quite sincerely the problem of which one do you value the best and enjoy the most, and that's not so simple.

Sue **And are you going to venture an answer on that one?**

Francisco Well it's more than an answer—I guess it's a statement about where I'm at. Sometimes I ask myself 'Why do you bother so much? Why are you trying to push this thing so hard?' You know, I could easily spend most of my time in the south of France in a beautiful little stone house that I have, just being it—just enjoying it. But then I will only be good as a subject and not first as a scientist.

Max Velmans

The universe has different views of itself through you and me

Born in Amsterdam (1942), Max Velmans did his first degree in Electrical Engineering at Sydney and his PhD in Psychology at the University of London. He describes his interests as folk guitar, sailing, and the nature of the universe. His work aims to integrate philosophy, neuropsychology, and mind-body relationships in clinical practice, to develop a programme for a nonreductionist science of consciousness. He is Professor of Psychology at Goldsmiths College, University of London. He has edited collections of papers on consciousness and is author of *Understanding Consciousness* (2000) and *How Could Conscious Experiences Affect Brains?* (2003).

Sue   What is consciousness, as far as you're concerned?

Max   There's a lot of confusion about that question, and I think it's really important to be clear about what's required for a definition. When you try to define consciousness you have to start in the right place, and then your understanding of consciousness gradually develops. The right place to start for me is the everyday experience that you and I have—for me in this particular moment my conscious experience consists of this three-dimensional room in which we're both sitting, you looking at me nodding and smiling, the sound of my voice, my hands waving in the air and so on. I might have some feelings, some images and so on, I might even have a few thoughts, but the bulk of what I experience at this moment is a three-dimensional phenomenal world extended in space.

It's really important to say this, because in my view most of the great debates about the nature of consciousness start in the wrong place: they are either explicitly dualist in their idea of what consciousness is, or implicitly dualist. And what I mean by that is that they start with an idea that consciousness is somewhere in the mind—a Descartian idea if you like, that there's something about me that you can't see which is above and beyond my body and brain, which is not located in space at all. That's the classical dualist Descartian notion that's explicitly dualist.

What's implicitly dualist is the reductionist reaction to that, which is to say: look, consciousness is something very ineffable and mysterious—we can't fit that into a natural-science view of the world, so we have to demonstrate one way or another, by hook or by crook, that this ineffable conscious entity is nothing more than a state or function of the brain.

Sue So let me get this straight: you think that most people reject the classic dualism of Descartes, with its complete separation between mind and body, but a lot of them, even though they've rejected Cartesian dualism, are getting into as much trouble in another way; is that what you're saying?

Max Yes, that's exactly what I'm saying, and the reason they're still in a lot of trouble is because there are some things that are shared, in the conventional scientific world view, with the original dualist model. In other words, you've got rid of the cloud floating above the brain; you've, so to speak collapsed it into the brain; but you're still dealing with a tacit idea of consciousness which doesn't resemble our ordinary experience.

Let me explain: normally we think that there's an external world around us that we see, which is physical, and from which we pick up energies through our visual systems and other sense organs. Our brain processes those things, and if you're a reductionist, you wind up with a conscious experience of the physical world somewhere in your brain. The only problem with that is that although it's absolutely true that the brain is deeply implicated in the necessary and sufficient conditions for having the kinds of experiences that we do, the actual resulting experience is not an experience in the brain; it's a three-dimensional phenomenal world extended around our bodies—like this particular moment, with you sitting out there in space.

So my subjective world—and it is my subjective world that I'm describing when I'm describing my conscious experience—is indeed what we normally take it to be—what we normally think of as the physical world that surrounds our bodies. So there never really was a split between the world as experienced around us and our experiences of it. Phenomenologically they're the same.

Now that's not to say that there isn't a world as described by physics, which is quite different from what we normally call the physical world. But in the way I look at it, the world as we perceive it is one representation produced by processes in the brain interacting with real energies in the world, which model the nature of the realities in which we're embedded.

Sue But you seem there to be describing the hard problem. You're saying there's this three-dimensional experienced world we're sitting in now, and you say it has a lot to do with the brain. But how come an experienced world with all its qualities of touch and feel comes about from stuff going on—objective physics stuff—going on in the brain? Are you saying you have some solution to the hard problem, to this whole great mystery?

Max Ah, there's more than one hard problem. Let's talk about the one that you raise, which is that brain states look like one kind of thing, and phenomenal worlds seem to be a different kind of thing; how do we construe the relationship between them?

If you accept, as I would, that neural causal processes in the brain in a sense produce these experiences, and that indeed there might be neural correlates in the brain going on at the very same time as you're having those experiences, but that there is something deeply mysterious about the fact that these neural states seem to be completely different from these phenomenal worlds—then, what kind of explanations would start to count as explanations of what's going on?

Let's talk about the relationship between electricity and magnetism. As far as we know, electricity is produced by electrons flowing down wires; but magnetism is represented as a field around the wire. You might say, 'That's very odd! How could something going down a wire, which we think of in terms of electrons, produce something which is actually outside the wire and described as a field?' The fact that things seem in the first instance to be different kinds of thing doesn't mean that if you understand them more deeply there might not be an understandable and deep causal interaction between them. So I'm going to give you a very simple scenario, a thought experiment.

Let's say that in some future state of neurophysiology we had actually isolated the precise neural correlates of a given experience; let's say it's a really simple experience, like this tape recorder in front of us, and you're the subject looking at the tape recorder, and I'm the experimenter inspecting your brain. Now the question that we're trying to get a handle on is, how does what I see in your brain relate to what you see out in space?

Sue Yes, this seems to me to get right at it: for me this is a private experience of the tape recorder there, and for you as the scientist, you're looking in my brain and seeing objective things going on there; is that what you're getting at?

Max Well, it certainly is the way we conventionally talk about it, but I want to challenge the point that you've just raised. I would put it to you that the actual truth of what's going on is that when you're looking at the tape recorder and reporting your conscious experiences, you simply report what you see, which is presumably a tape recorder out there in space; and when I'm looking at your brain, at the neural correlates of your experience, I'm simply reporting what I see.

At first glance it's not obvious that what I see has any more objective ontological status—more objectivity—than what you see. One of the ways to see that is that our roles are interchangeable: I could now look at the tape recorder instead of looking at your brain, and instead of looking at the tape recorder you could look at my brain. That makes you the scientist and me the subject. But why are my experiences of the tape recorder suddenly subjective, when, with my scientist hat on, my experiences of your brain were objective? Clearly the objective-subjective relationship has to be thought of in a deeper sort of way, because that kind of objective-subjective switching doesn't make sense.

Sue But that kind of implies that we're wrong in separating out the subjective and the objective, and that indeed when we're doing almost any kind of science we're doing the same thing.

Max That's right. I would argue, for example, that there are four kinds of objectivity, and that they tend to get confused. You can be objective in science in the sense of making intersubjectively validated observations—so you and I can agree between us about the nature of the tape recorder out there in the world. The second kind of objectivity is being dispassionate, trying to be truthful, not cooking the books, not letting wish-fulfilment enter into your data entry or your analysis of the results. And the third kind of objectivity is making your procedures sufficiently explicit and detailed so that anybody else carrying out those procedures could carry out the experiment in the same way. But here's the rub: the one kind of objectivity you can't have is to make an observation that is objective in the sense of being observer-free, an observation that doesn't somehow involve the experiences of the observer. That's not possible.

Sue So we have a situation in which all of us as scientists are implicated in the science we do, and can't make completely objective measurements; but nevertheless I get this feeling that we do believe that there is a real world which our experiences are experiences of; and as soon as we say that we're back into the problem. Are you really getting out of this somehow, and I haven't understood how?

Max The position I want to defend would be called, within philosophy, critical realism. In spite of the fact that I would regard the world that I experience around me as my experience, and also call it my physical world, I nevertheless agree with you that the sensible position to take is that that phenomenal representation which I'm the focus of, which is all seen from my perspective, is a representation of something which is autonomously existing.

Sue Then you are a kind of a dualist!

Max Only in a sense. And it's not dualism at all in the old sense of there being two kinds of substance in the universe. In fact ultimately I developed a position which I call reflexive monism.

Sue Perhaps you could explain that as simply as you can.

Max OK, let's say you go with a big-bang theory of the universe: in the beginning all of us, all our bodies, all our matter, all our potential experiences, all our potential thoughts, everything that could possibly exist about us is packed into this tiny, tiny mass of infinite density, and it explodes. Then the universe expands, gradually differentiates, and eventually, on planets like ours, organisms evolve which are further higher-order differentiations. But of course we know from our own lives that one aspect of our being isn't just the fact that we're walking around in separate bodies; each of us from our own perspective also has a view of the whole thing. So there is a sense in which the whole universe is differentiated into bits which have this wonderful ability to have a view of the whole; and that, in a few words, is reflexive monism.

Sue So you're imagining a universe out of which pop up, as it were, centres of viewpoint, places from which there is a viewpoint; and these, I suppose, have to be complex information processing systems. But then you come to the big question, what kind of stuff do you have to have in these bits of the universe for it to have this self-reflexive quality?

Max Absolutely. And this is a long story.

Sue Can you make it a short story?

Max I can try. There are two fundamental positions you can take on what needs to happen: discontinuity theory and continuity theory. There are many discontinuity theories, which basically say that the universe developed from some totally non-conscious insensate mass, and then at point x the light suddenly switched on. The view you've put would be one popular version of discontinuity theory; it would say, for example, that when biological organisms developed to a certain state of complexity, or, if you like, when their brains attained a particular level of complexity, suddenly the lights switched on. Now of course all discontinuity theories have a problem, because whatever theory you happen to have you can always scratch your head and say, 'But why did the light switch on when *that* happened?'

Most of us for example would be happy to say that our chimpanzee cousins were conscious; any of us who have got dogs and cats would be pretty clear that they are conscious; so what about frogs? You might be willing to accept this and say, 'Maybe any creatures that learn have consciousness.' Then of course you have to deal with the fact that a lot of our learning is unconscious anyway. So all discontinuity theories have a big problem: Why did *that* change in structure or in functioning switch the lights on?

Now the alternative is continuity theory. Continuity theory says: consciousness is fundamental in some way, although our particular form of consciousness exists only with our particular biological forms: our senses, social structures, languages, and so on. There might be a fundamental relationship between consciousness and matter—one version of this is panpsychism—and what happens as evolution progresses is that the forms of consciousness coevolve with the evolution of the forms of matter; so there never was a point at which consciousness first switched on.

In my own work I remain neutral about which of these is the truth, but if I had to make a bet in terms of which is the more elegant theory then I'd have to say that I find continuity theory more attractive.

Sue So you're at least tempted by the theory that consciousness is in some way fundamental in the universe, and that the particular sort of consciousness we humans have emerges because of the particular kind of brains that we've got.

Max Absolutely.

Sue This sounds somewhat similar to David Chalmers' theory; is it—or if not, how does your view relate to his?

Max On that issue there is a specific difference. Dave Chalmers also gives consciousness a very wide distribution, but the big difference between the story I've been telling and the story that he tells is that for him matter itself doesn't matter for consciousness, which is very extreme.

I actually call his position pan-psychofunctionalism, as opposed to panpsychism, because he suggests that the only necessary physical concomitant of consciousness is information, irrespective of how that system is embodied. So if we take a non-biological organism, let's say a robot that we construct to function just as we do, according to him, simply by virtue of functioning as we do it would experience as we do. Now I'm not so sure about that, because he's saying that the stuff we're made of makes no difference. That's a very strong claim and I simply wouldn't at this stage want to go down that route. It might be the case, for example, that silicon robots have no experiences; it might be the case that silicon robots have distinct silicon experiences; or it might be the case that silicon robots have human experiences—depending on whether only functioning matters or whether there's something about the combination of functioning and the material in which it is embodied that matters.

Sue So let me get this straight. As far as Dave Chalmers is concerned, information is the critical thing; information has two aspects to it, if you like: how it looks from the outside and something it's like to be the information. You're not following him down that track; you're saying you've got to have an information-processing system, made of some particular kind of matter.

Max Well, we've got to be careful about this, because actually Dave developed a theory which was very similar to one I produced in 1991. So I also have a dual-aspect theory of information, but it differs from Dave Chalmers', in very specific ways. Dave, for example—in my understanding anyway—has three things in his complete theory of mind and consciousness: firstly, a functioning system—and it's basically the information in the functioning system that matters; secondly, a natural fact about the world, that information is accompanied by phenomenal experience—he calls this naturalistic dualism; and then a set of bridging laws which connect the first with the second. Now, in many ways I want to say something very similar—and actually that's a convenient place to go back to a question that we left right at the beginning of the interview, which is, what's happening right at the interface of consciousness with the brain?

I would say that for every distinct experience there will be, if you like, a distinct correlated state of the brain that encodes identical information. So whenever you get discrimination in the phenomenal world you will also be able to see differences in the neural correlates. But what connects this neural state that I can see if I'm looking at your brain, with the phenomenal experience that you're having? I would say that in a rough way the situation is not unlike the sort of thing that occurs in quantum mechanics, where you find that if you try to give a complete description of something like an electron, the way the electron is described very much depends on the observational arrangements: in certain kinds of observational arrangements the electron simply looks like a wave and in other observational arrangements the electron simply looks like a particle. I think that there is a direct analogy with what's going on with conscious experiences and their neural correlates in the brain.

Sue I'm tempted to think this is a cop-out; on the one hand I think it might help, maybe all you're saying is that it depends where you're looking from, and where you're looking from will determine whether it's either a world of experience or bits of brains doing things; but then I think, no, this hasn't got at the heart of the problem because those are different kinds of things. Are you saying they're not; am I missing this; am I failing to understand it?

Max You're not wrong, but the problem is I can only do soundbites here, so I'm giving you little bits of the story, and all these questions...

Sue Yes but I need to push you to that.

Max OK, which bit of the story do you really want to push me to?

Sue All right, I want to understand whether you think that subjective experiences and objective brain activity are just two aspects of the same thing, and that they depend on where you're looking from. Is that what you're saying?

Max Yes, the two aspects of this information are being displayed, if you like, in two different ways. The claim that I'm making is that you could have identical information, which, depending on how you display it or view it or hook into it, might actually be manifested in completely different apparent forms.

Sue That makes it clearer. Now I want to ask you about zombies; in the classic zombie thought-experiment you have this creature which can do everything that we can do, let's say it's a Max Velmans lookalike and it does everything you do, but it doesn't have any inner experience. Now from everything you've said, and based on your theory about consciousness and its relation to the brain, do you believe that in principle zombies are possible?

Max Well, I think you've got to distinguish what you might call logical possibilities from actual possibilities.

Sue Logical—I'm not talking about actually here in this world. I mean, in principle, given what you've said about the way you understand what consciousness is—in principle would that allow the possibility of a zombie?

Max Well, well I have to insist on going back to the distinction because…

Sue All right, go on then…

Max The reason I have to insist on it is because it really matters: I think you can only get certain kind of change out of an argument based on logical possibilities. You might indeed say, because it's conceivable that you might have a creature…

Sue I explicitly ruled that out in my question; I'm not interested in whether it's conceivable but whether it's possible.

Max No, OK. I want to get on to the real issue in a minute, but the fact that a zombie is conceivable is important, because it means that once you've conceptualized the nature of a working brain system, there's still some work to do to connect that, at least conceptually, to the phenomenal experience. If zombies are conceivable, it isn't automatic that once you know all about the brain state you also know all about the experience. But I agree with you that the much more important question is what's actually possible; and, given the universe that we know, if you absolutely replicated the functional and the structural conditions in our brains, and actually had an artificial brain functioning just as ours does, I think I personally would rule out the possibility that it didn't experience as we do.

Sue So your answer is no, you don't believe in the theoretical possibility of zombies?

Max Well, as I say, I can conceive of them in some universe where the laws of nature are completely different...

Sue Do you have free will?

Max Do you want the proper answer?

Sue No. I want the instant thought.

Max Yes or no, yes or no? OK, the quick answer is that my sense of being free is, I think, a genuine sense. That's not in any way to argue against determinism in science, but I am the kind of creature that's capable of choices—I can do what I want. But I can't want what I want, so there are deep inbuilt constraints. Yet there is a range of activity within which I can do the things that any cognitive psychologist would accept I can do: we can attend to certain things rather than others depending on what interests us or is important to us; we can make decisions about what we're going to do on the basis of the things we attend to. And there are many ways in which our minds are sufficiently complex to make the choices that we are required to make, and then of course to take responsibility for the choices that we do make.

Sue Right, I want to ask you one more question. Doing work on consciousness is in a sense working on yourself, because this consciousness now is where you're starting from; do you find that doing this kind of work has changed your consciousness or changed your life in any important way?

Max Sure. I think this is a really important business because the theories that we have about ourselves or about the nature of the world become frameworks within which we live. And so they constrain what we think is possible, what we think is real; so it really matters.

For example, if it turned out to be the case that materialist reductionism is true, and all our experiences were entirely epiphenomenal, of no consequence to anything, that would be a deep problem for most of us. Now you might say 'Well if that's the truth, why should it bother us?' But I would claim that's not actually the truth—that neither logically or scientifically can you make that claim stick.

My reflexive monism gives me a more expanded version of the reality of my own conscious being, of the fact that I, as a being, am in some sense implicated in constructing the world that I experience. Admittedly the world exists whether I do or not, but there's much more a sense of being embedded in reality in this way of looking at things. The first person perspective, which is the perspective with which I live my life, is given its proper status, without denying any of the scientific facts which are just as important as they always were. For me it's a fuller vision of the universe and a much more comfortable one to live in.

Sue One of the things that I try to do is to hone my own experience—to really look at what our experience is like—because it's so easy to take for granted. That's why I meditate and do other things to sort of get at the directness of experience; do you do anything like that?

Max Yeah, I think that once you start taking experience seriously and realize that everything that you try to know about the world starts with your experience, then you have licence, in a way, to explore your own experience. In the end, for me, the science of consciousness would include the whole business of listening to the messages that we get from people around the world who've tried to change their experiences, sometimes seemingly with beneficial effects, and picking up the odd pearl of wisdom where we find it. So let's say you're meditating, and you no doubt were led into that by somebody who seemed to embody qualities that seemed to be rather nice to have; in a sense that provides a kind of empirical test, that engaging in certain kinds of procedure might produce beneficial results. For me that's just science; that's the empirical method being applied, in this case to possible methods for changing one's own inner states.

Sue Have you had uncomfortable experiences in your journeys? I presume you've journeyed through believing in different theories of consciousness over the years you've been studying it; have you ever had real crises or traumas, thinking, 'God, if it's like this, I can't cope with it'; has it really touched you that deeply at any point?

Max Well, it is a deep thing, but obviously when you produce a different theory of the sort that I have, the first thing that occurs to you is, 'I must be mad', and the second thing that occurs to you is, 'I'm probably wrong'; and the third thing is, 'Somebody's probably said it before'; and so it took a long time for me to work around the theory, debate the theory, discuss the theory, try and write about the theory, and so on, before I convinced myself that there weren't any obvious gaps. Now of course there *might* be gaps. But at the moment it's not clear to me that there are any; at any rate nobody has clearly and explicitly pointed them out to me.

It's always a partial theory of course; there's no question that my work's just a little bit of a much bigger picture; but apart from those kinds of anxieties I personally find the reflexive approach a much more comfortable way of conceptualizing my own nature.

I think in many ways though, that you and I are the same: I was always the critical sceptic, so to speak—the intellectual type who wouldn't believe anything anybody told me; but because this way of looking at things logically coheres for me, and it's not actually inconsistent with the scientific evidence, and it's also consistent with my everyday experience, I find I'm rather supported by this structure, more than finding it troublesome, worrisome, or anxiety-provoking.

Sue Is there any sense in which you think you have come to this theory because you like it; do you think you've avoided any other theories because they really are too uncomfortable to live with?

Max No, I don't think so; but there's an interesting and deeper issue here, which is that there's a kind of deep pattern-recognition that we have: there *is* a reality and we *do* embody it—we don't understand it, but we can recognize when somebody's talking rubbish or not. For example, I, like you, have had to grow up with behaviourism and all sorts of -isms—and I thought, 'There's no way I can believe any of this stuff.' But I really didn't get into this approach because I wanted things to be like this; there's no way that I could even conceive of consciousness like this if I started from that point. It was just a matter of following my nose and thinking, 'Hang on, I'm winding up in a different place here.'

It's always tentative of course; I have a bit of scepticism, certainly about the theory's completeness—I know it's not complete, and there is always the possibility that it might be fundamentally wrong. It might be, but it's not obvious to me that it is.

Daniel Wegner

*Don't think about
a white bear*

Born in Canada (1948), Daniel Wegner studied physics in Michigan but
changed to psychology as an anti-war statement in 1969, and began work on
questions of self control, agency, and free will. He has done numerous experiments on thought suppression, as well as how the illusion of free will is created. He not only plays the piano but has four synthesizers and composes
techno music. He spent 15 years teaching at Trinity College in Texas, and then
became Professor of Psychology at Harvard University. He is author or editor of books on the self and social cognition as well as *White Bears and Other
Unwanted Thoughts* (1989) and *The Illusion of Conscious Will* (2002).

Sue We have this new field called consciousness studies; there are people
from neuroscience, from philosophy, and so on, all agonizing about the
problem of consciousness; what's the problem?

Dan I think the main problem is that everyone has a consciousness but
they have no access whatsoever to anyone else's. This is the problem
of other minds and so far it's insurmountable: we don't know what
it's like to be someone else, and we don't know what it's like to have
another consciousness.

Sue Some people have argued that because we can't access anyone else's
consciousness the whole idea is incoherent.

I think there must be someone it's like to be you, sitting there, and I believe that there's something it's like to be me; but some philosophers would say that that's incoherent, that there isn't anything that it's like to be you or me.

Dan The problem is that each of us is the only thing we will ever know what it's like to be.

Sue That's kind of scary, isn't it?

Dan Yes. And anything we do has to be a matter of inference, rather than actually being the other person. So the question is, how do we go about appreciating that? What are the signs of another person's consciousness? What could lead us to experience life the way they do? Basically it's what other people tell us.

Sue Is this process of inferring what it's like to be somebody else a different kind of inferring from what is done in physics, when we infer things about sub-atomic particles from remote machines that give read-outs? Is it fundamentally different to be dealing with first person subjectivity?

Dan I think so. The deep problem of psychology, and actually of the social sciences generally, is that they're at once studies of an object in the sense that physics is the study of an object, and studies of a subject, of what it's like to be the person who is being studied. And unfortunately the scientists who are doing the studying of consciousness are both the objects of study and the subjects, so it all becomes much more confusing than the other sciences. You need somehow to be objective about subjectivity, which in a way is the deepest conundrum we can think of.

Sue When you say that we are both the subject and the object of what we're doing, doesn't this mean that potentially we must change ourselves in the process?

Dan There's certainly a lot of worry and discussion about that. As I see it, the field of consciousness studies right now is made up of a large group of people who are particularly concerned about becoming objective, as objective as one could possibly be; and then another large group of people who are completely given over to subjectivity; who want to talk about experience, and about what it's like to be human, and what the world looks like, and how things seem, and how it all works in their own minds. That area has classically been known in psychology as phenomenology, and it's come in and out of the science of psychology. There have to be ways of building bridges.

Sue Do you think we are at least beginning to build those bridges?

Dan Oh, certainly. Many of the exciting things that are going on in this field have to do with that bridge. Take the work on phantom limbs being done by Ramachandran. He finds that if you have a visual representation of another hand in the position where yours ought to be, you may experience the movement of that hand as though it were your own. So if you look at, for example, the reflection of your left arm in the position where your right arm was once before, and that left arm moves in the mirror, you may actually experience it as your right arm moving in the space where there is in fact no physical arm.

Sue This is kind of creepy, because if you can feel that non-existent arm just as much as your real arm, it suggests that your real arm in normal consciousness might be some kind of illusion, doesn't it?

Dan I prefer the term construction to illusion, in the sense that we have to build an overall idea of what our body is and what it's doing. Somehow there's a way that all this gets projected into consciousness; there has to be a mechanism. There's someone in the projection booth producing all of this stuff for us, and obviously those mechanisms are the key thing that we're interested in finding out about.

Sue But isn't this a completely false analogy, or metaphor, this idea of the projection booth? It sounds similar to Dan Dennett's Cartesian theatre, in which we sit, somewhere in the middle, looking out at the world, or imagining things as though on a mental screen. But we know that the brain isn't like that; the brain's just neurons; so how can you make sense of a metaphor like that?

Dan Well, there doesn't have to be a place in the brain where this projection occurs. The experience doesn't need to map perfectly onto an array of spots in the brain which is the projection area. This is, I think, one of the big puzzles that people are working with in this field—how the projection takes place; but it certainly is true that all of us experience the world as a rich field of perceptions and events and we need to understand what that's like. I'm not sure that I'm quite in agreement with the way Dennett tries to undermine the idea of a personal phenomenology.

Sue Isn't it more that he's trying to undermine the idea that there is an audience in there, watching the stream of experiences go by?

Dan The magical thing is, it seems as though there's a self watching these experiences.

Sue Right, so where does this magic come from?

Dan How does it get constructed? That's a great question.

Sue Well?

Dan There are clues here and there. For example, new selves seem to be constructed in cases of dissociative identity disorder when people develop new multiple personalities, and alternate selves arise in cases of apparent spirit possession. Processes like the ones that create such selves may be responsible for creating each of our initial selves. Subjective worlds may be created, not born.

Sue I know you've been doing work for some time on thought suppression. Can you explain what that means?

Dan Just trying not to think about things, whether that works and how people do it.

Sue It seems extraordinary how difficult it is. We have this illusion that we're in charge of what we think about—but actually all these thoughts just seem to come, don't they?

Dan Right. We did some little experiments where you ask people to try not to think about a white bear, and then to speak out loud into a microphone as they're trying not to; and they mention it about once a minute—even when they try for half an hour, it just doesn't go away.

Sue How accurate can such a method be at getting at consciousness? I mean, if you tell somebody to talk aloud, it's not at all similar to just sitting thinking.

Dan Well, you can get at it without doing reports. One way is to ask people to talk about something that's emotional; and of course you get an emotional response, you find that their skin conductance level goes up. Then you ask them to try not to think about it, and they talk about whatever else—and meanwhile their skin conductance level also goes up.

Trying not to think about something creates this ironic process where that thing automatically comes back to mind. People do it all the time, and it's probably the beginning of a whole lot of mental turmoil and psychopathology. If somebody tries not to think about something that makes them anxious, what happens is that it blows up in their mind and becomes more accessible all the time.

Sue And yet it can't all be pathological. We have to pay attention to the things that are important to us at the time; we can't go around thinking about all these annoying things that might upset us; we've got to suppress them, haven't we? Or at least push them away?

Dan We postpone thoughts all the time: a thought comes to mind that, 'I need to do this before I go on this trip, and but I can't do it right now, so I'll just put it off'; and then it keeps popping back until you do it, so it acts as a kind of a little internal alarm that reminds you of the thing you've postponed. The trouble is you can't postpone forever, and that's what thought suppression is. It's the desire to keep something out of mind *from now on*. And so it continues to remind you; it's always there.

Sue So how do you think a healthy person copes with the problem that there is too much to think about all the time?

Dan People find new things to think about. There's a very subtle difference between trying not to think of x and trying to think of y, and many people just go for trying not to think of x, and don't realize that if you just wander off into a whole new domain you may not worry about x any more.

Sue But if you're right it will come back in some other way. Even though you think about y instead, x is still lurking there with its emotional connotations. Wouldn't a more healthy way be to at least give time for x to come back and be dealt with?

Dan That's another very important technique and is in fact the basis of psychotherapy: you go ahead and talk about your problem. People find a lot of peace in expressing the things that they're most afraid to talk about; if you go ahead and chat about it with your best friend, or your therapist or your minister or some confidant, or even just write it down at length for yourself, so that you've thought it through, that dispels this need to suppress, and makes the whole thought easier to deal with in the future.

Sue But I thought there was evidence from psychotherapy that going over all these emotionally arousing things can actually make you more angry, or more upset, instead of helping.

Dan That can happen; I think it's possible to dwell too much on something. I guess there's a happy medium of thinking about it to the point where you've figured it out, and can move on. Jamie Pennebaker's research suggests that expressing thoughts can help us get new perspectives on them and help us to dispel the thoughts from our minds.

Sue **So from these studies of thought suppression, what sort of picture do you have of an ordinary person's stream of consciousness?**

Dan I like to think of the example of a school bus. There are lots of kids on the bus and they're all running forward and saying, 'Is it time for me to get off yet?', and the bus driver has to hold them back. These are all of the thoughts we have every day—there's a bunch of them in the back of the bus, and they all want to pop up into the front, and be let off into consciousness, and the fact of suppressing keeps them on the bus. They're always running to the front, and they become more and more annoying as a result of being held back. If you just let them off then it's over, they're done, and they wander away.

Sue **Could we reasonably think of these as memes trying to get space in our brains?**

Dan As I was reading your book I had the thought repeatedly that some of the most powerful memes are not memes that everybody thinks and talks about, but the ones we specifically avoid. There are thoughts that you don't want to have that end up coming through in the things you say and talk about with other people, and that then become unwanted thoughts for them. I think society has a lot of unwanted thoughts that are transferred from one person to another by this desire for avoidance.

Sue **Do you think we have free will?**

Dan It certainly seems as though I do. My work these days is concerned with the feeling of freely acting. I'm trying to understand how that feeling comes about, because it's not part and parcel of action; there are lots of actions that look exactly the same as the actions that are done with the feeling of conscious will, and yet they don't feel willed.

Let me give you some examples. There's a set of behaviours that we refer to as automatisms, which probably were best known as parlour tricks in the spiritualist tradition 100 years ago, things like Ouija boards and automatic writing. Table-turning is another favourite of mine: people sit around a table waiting for a spirit to move it, and very often you'll find that within some minutes the table will start to move around the room. Dowsing is another example; people feel that the divining rod is moved by some force towards the earth as they're walking around with it; they don't feel that it's voluntary at all, but in fact the action appears to be perfectly voluntary. I'm not willing to try to test the hypothesis that spirits are moving it; I'm much more interested in why it feels involuntary to the person who has clearly done it.

Sue Well, you got there a lot more quickly than I did! I spent years and years and years trying to find out whether spirits were moving things, before I finally concluded there must be some psychology here, rather than parapsychology.

Dan The theory I have is that the mind produces actions for us, and it also produces thoughts about those actions. We feel will because we see a causal connection between the thoughts and the actions. Sometimes the thoughts don't get there quite in time to precede the action, or the thoughts are attributed to someone else, as in the case of the Ouija board. So we end up losing that feeling of will.

Sue Let me get this straight. Are you saying that in our normal life we think we're going to do something, and then we do it, and we say 'Oh, that means my thoughts caused it'; whereas really it's something like this: there's some sort of underlying brain process that simultaneously causes our awareness of an intention and also the action, and we end up thinking there's a causal relationship where there isn't?

Dan That's put very nicely, yes.

Sue So how can you test this theory? It sounds very good in practice, but surely it's rather difficult to get at it?

Dan One way to test it is to cause people to perform actions that they didn't do on purpose, and simultaneously to provide them with thoughts of what the action will be, and see if they experience will as a result.

Two students in my laboratory, Betsy Sparrow and Lee Weinerman, arranged for people to perform a pantomime called 'helping hands'— I think this was also in an old Marx Brothers movie—where one person approaches another person from behind and puts their hands under the first person's arms so that this second person's arms are coming out in front. So now it looks like the person in front is moving their hands. The person in back also puts on gloves and a kaftan so it's not clear whose hands are whose, and the subject watches those arms in a mirror. We instruct the person in back to move their hands around and clap a few times and touch the person on the nose and play catch with a little ball and things like that. And you ask the subject, 'Does this feel as though these hands are yours, and that you are consciously willing their movement?'—and they normally say, 'No, it's a cute illusion but it doesn't really feel like they're mine.' But if you play them an audio tape of the instructions that are being given to the person in back, saying 'Now clap three times, now touch your nose with your right hand,' and so forth, they're much more likely to say, 'Yes, it feels like I'm doing this. I know at some level of course they aren't mine, but I get this funny sense that these are things I'm doing.'

Sue **So the implication is that in normal life if I think, 'Clap three times,' and then these hands do it, I infer that the thought caused the clapping, even when actually it was some underlying brain mechanism that caused them both.**

Dan Exactly. And the result is that I feel that I've willed this. I think of the feeling of will as something like an emotion: it surges forward; it labels the experience as yours; it authenticates it. I don't think it's a rational process of figuring out what you've done; it occurs almost as a rush of recognition, 'There, I've done it again; I'm clapping three times.'

There's another series of studies which Thalia Wheatley and I have done, based on the idea of the Ouija board. We have a participant in the experiment put their hands on a little board that's resting on top of a computer mouse, and the mouse moves a cursor around on a screen. The screen has a variety of different objects, pictures from the book *I-Spy*—in this case little plastic toys. Also in the room is our confederate; both of them have headphones on, and together they are asked to move the cursor around the screen and rest on an object every few seconds, whenever music comes on.

Sue **So they've both got their hands on this equivalent of a Ouija board...**

Dan Right, and they're both moving together. Most of the time they hear sounds over the headphones they're wearing, and some of these are names of things on the screen. The key part of the experiment occurs when, in some trials, the confederate is asked to force our subject to land the cursor on a particular object, so the person who we're testing hasn't done it, but has been forced. It's just as though someone was cheating on the Ouija board. We play the name of the object to our participant at some interval of time before or after they're forced to move, and we find that if we play the name of the object just a second before they're forced to move to it, they report having done it intentionally; if we play the name of the object well in advance—some 30 seconds before—they don't get that experience; and if we play the name of the object after they have reached it, they don't get that experience.

Sue So the feeling of having done something comes about not because you really have done it, but because there's a short gap between thinking about something and its happening. Does this mean that the feeling of agency doesn't prove that there is real agency?

Dan Yes, the feeling of agency can be fooled—and yet, we go about our daily lives feeling the opposite: we have the intuition that our feeling of agency is proof that our minds are working that way. In fact we're not that insightful about our own mental processes.

Sue I've met lots of people who claim to be able to move the clouds around, or make lights come on and off in the street; is this the same effect?

Dan Exactly.

Sue And what do you think is its function?

Dan Oh, there are lots of functions. I think the most important function is establishing who did what. You can think of life as a big whodunnit in which we're all concerned with whether particular actions were done by us or by someone else. If we have this feeling that comes forward every time we do something, or infer that we've done something, it acts as a way of labelling things as our actions. That way we can feel responsible for them; and we can morally judge people who have done good and bad things. We're willing to put people in prison for actions if they feel that they did them, and oftentimes will put them in psychiatric treatment instead, if they didn't feel that they did them. We make a very strong distinction in the law between actions that people feel responsible for and that were intended, and those that were not, and I think it's because of this preview system that provides our intentions, and this sense of authorship that we each have as a result. We trust each others' sense of authorship, and use it as a way of allocating punishment and rewards to people in everyday life.

Sue I can see how important that is, but it's rather worrying in a sense, because we are putting the weight of all these important legal decisions on an inference that isn't always correct.

Dan True. Well, nobody said people were perfect. This is a very nicely oiled, well-running guessing system that sometimes goes wrong;

when it goes wrong we end up with automatisms like the Ouija board or automatic writing.

There are also some cases of hypnosis that I think this would be a good way of understanding: when a person is hypnotized often they do things that appear from the outside to be perfectly voluntary, but that the person experiences as being totally involuntary. Hypnosis, then, would be a system in which we're undermining the normal process for inferring our own conscious will.

Let's think of it this way: each of us has a mind that produces for us a sense of virtual agency, of feeling that we are a self who does things; this ends up being a very useful accounting system and a useful way of keeping apprised of our actions as opposed to those of others, or of the world. To say that it is a virtual system doesn't mean that it's any less real, if it in fact ends up guiding subsequent behaviour. So it's very important, even though it's a construction, as opposed to a reality.

Sue But if I ask you 'Can a thought cause an action?', what would be your answer?

Dan I'd be perfectly happy to say that that's possible; and in fact I think that's an important finding of much of cognitive psychology—that thoughts do cause actions. The fact is, though, that consciousness doesn't always know that if a thought has caused an action it should create an experience of will associated with that.

Sue But how can a thought cause an action? I'm talking about conscious thoughts—we have a subjective private experience of thinking, 'I am going to touch my nose.' How can that subjective experience *cause* something like a movement of the hand which is a physical and objective thing?

Dan I'm not sure that I would say the subjective thing causes the objective thing, as much as saying that the subjective experience is one of the indicators we have of the objective system.

I'd like to think that most of the time the subjective feeling is riding along; you might think of it as the mind's compass, that gives a sense of where the body is going, and we're watching the whole process go on. So it's not as though subjective experience is something that will never matter, it's just that at the time of the behaviour it's a view of what's happening, not the initiator of what's happening.

Sue You said 'most of the time': are you trying to allow here a little loophole for real effects of subjective thoughts on the world?

Dan I don't think so. I think that might just have been my way of trying to be nice to those people who would like to have a subjective driver at the wheel as they career through life.

Sue It's very natural and understandable to want to feel that I am the driver—that there is somebody running my life, and living it—and yet what we're learning about the brain suggests that this isn't true. Shouldn't this affect the way we live and the way we feel about ourselves?

Dan I'm not sure that we're at the point in the scientific study of this that we need all start behaving differently; it's not clear to me that I would behave differently as a result of what I know, and until I reach that kind of personal determination I'm not ready to recommend something like this to anyone.

Sue So this work you've been doing on thought intrusion, on the sense of willing actions, hasn't affected the way you live your life?

Dan I would have to say that it gives me a sense of peace. There are a whole lot of things that I don't have to worry about controlling because I know that I'm really just a little window on a lovely machinery that's doing lots of things. It also gives, not so much a sense of inevitability, but perhaps a sense of correctness to the behaviours I do—that not all of them have to be chosen; I don't have to worry about every little thing; things will happen well, and have happened well throughout my life, as a result of simply allowing this machinery to do its operation. I was recently faced with a major life decision, and part of the process of deciding in advance was the knowledge that after I'd made the decision there might be a period of regret but then I could start looking forward to things falling into place, that I would decide that I had done the right thing, and that people around me would help me continue to believe that I had done the right thing.

You know the basis of many religions of the world is the peace that comes from not feeling in control—being able to give away control to your god.

Sue Yet there's a difference between relinquishing the control to a god, and—as in the case of neuroscientists, most of whom don't believe in a god—relinquishing it to the world. It's more like giving oneself up to the universe than having somebody else in charge.

Dan I guess that's another name for a god.

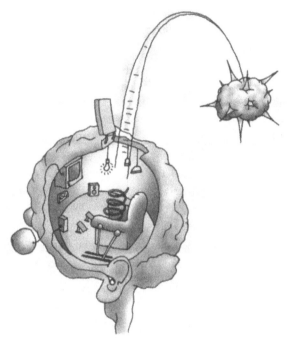

It's time to jump out of the Cartesian theatre and relinquish myself to the universe.

Sue **So now that you have, not an ultimate but a somewhat deeper understanding of these processes, would you say that free will as we normally think of it is an illusion?**

Dan Yes, it is an illusion, but one with what you might call a 'bottom'. It feels very real. The experience of conscious will happens not just to the mind but to the body, providing a kind of 'authorship emotion' that highlights for each of us what we feel that we've done.

Sue **You recently divided your fellow scientists into robo-geeks and bad scientists; what did you mean by that?**

Dan It's the distinction between people who feel that behaviour is controlled by mechanisms and those who feel that they consciously chose what they do. In jest I called the mechanistic types the robo-geeks and the other group the bad scientists. The robo-geeks would be the folks who are completely committed to the idea that we're going to make an objective study of humans and be able to understand them as mechanisms. The bad scientists would be the people who preserve some sense that their conscious will is an authentic experience of what's going on in their minds, that their consciousness does produce their activity. And in the field of psychology, my field, we're about equally divided.

Sue It's slightly cruel of you, isn't it? Are you really saying that you can't be a good scientist unless you believe that you are a robo-geek?

Dan I labelled them by the worst that each group thinks of the other. So I think the robo-geeks would point to the others and say they're bad scientists for not believing in mechanism, and of course the people who believe in free will think these other folks have somehow descended into the realm of robotics.

Sue Well, I'm definitely a robo-geek.

It seems to me that, if you reject free will, there are at least two ways you can go: one is to say that all the decisions that this body takes are going to be made anyway, so it would be silly to have any sense of willing them, and I should just drop that sense of willing and live completely without making choices. That's what I do and I don't think you become less human, or lose all the richness of life by doing so.

An alternative is to say, 'I know that really it's all mechanisms but I will just live *as if* I'm really doing it,' knowing in the back of my mind that I'm not. Do you think it matters which you do? And which do you do?

Dan I do the 'as if'. And I think almost everybody who's happy and healthy tends to do that.

Sue Oh dear.

Dan Imagine riding around a very complicated robot that has billions of circuits inside that's doing all kinds of interesting things. There's a certain enjoyment that comes from knowing what it's going to do next—even if there was something in the robot saying, 'We're going over to the left now. We're gonna climb up that hill.' That's really the position we're in with our own minds.

It's not a sad illusion that we have conscious will; it feels good to think we know what we're doing.

Glossary

Please note that this glossary is not intended to cover every topic in consciousness studies but is meant as a simple and personal guide to some of the ideas that pop up, unexplained, in the conversations. For full glossaries and further information on such topics see:

S. J. Blackmore, *Consciousness: An Introduction* (London: Hodder & Stoughton; New York: Oxford University Press, 2003).

R. L. Gregory (ed.), *The Oxford Companion to the Mind* (Oxford: Oxford University Press, 2004).

R. A. Wilson and F. C. Keil (eds.), *The MIT Encyclopedia of the Cognitive Sciences* (Cambridge, Mass: MIT Press, 1999).

The Stanford Encyclopedia of Philosophy http://plato.stanford.edu/

For the articles and books from the *Journal of Consciousness Studies*, see http://www.imprint.co.uk/jcs.html

I have given one or two key references for those who wish to follow up any of the topics listed here, using those by my conversationalists when possible.

automatism

The term can be applied to any automatic behaviour, including sleep walking, but usually refers to automatic writing, or the use of the Ouija board or planchette to communicate with spirits. Wegner discusses automatisms in the context of how the sense of being responsible for one's own actions comes about.

Wegner, D., *The Illusion of Conscious Will* (Cambridge, Mass: MIT Press 2002).

binocular rivalry

When different images are shown to the two eyes they compete for dominance, that is, instead of the two pictures merging, they tend to alternate. The phenomenon has been researched since the late nineteenth century but only recently has the neural basis of rivalry been discovered by Logothetis and others. The effect is often described as though two stimuli are 'competing for consciousness' or competing to 'enter consciousness', but note that this way of thinking may imply a competition to enter a special place or process, or to be displayed to an inner observer— in other words, entry to a Cartesian theatre. See figure p. 70.

blindsight

When people suffer extensive damage to V1, the primary visual cortex, they are left with a scotoma; an area of the visual field in which they cannot see. In 1978 psychologist Lawrence Weiskrantz discovered that when he presented stimuli to a patient's blind area and asked him to guess its orientation, direction of movement, or other features, he could guess correctly most of the time. In other words the guesses revealed the use of visual information when the patient said he could see nothing. This paradoxical condition has been much disputed. Some people claim that it shows vision without awareness and is equivalent to partial zombiehood (implying that consciousness can be separated from function, or even located in a particular part of the brain). Others point out that there are many visual pathways and patients can use information from, for example, the fast movement system or eye movement system to make guesses, while being unable to see normally because the object recognition system is damaged. See figure p. 216.

Weiskrantz, L., *Consciousness Lost and Found: A Neuropsychological Exploration* (Oxford: Oxford University Press, 1997).

Kentridge, R. W. (ed.), (1999), 'Papers on blindsight', *Journal of Consciousness Studies*, **6**, 3–71.

brain imaging or brain scanning

There are now many methods of brain imaging including PET (Positron Emission Tomography), MRI, and fMRI (functional Magnetic Resonance Imaging). These are frequently used to study the neural correlates of consciousness, revealing which areas of the brain are more active when people report certain experiences. The problem lies in the interpretation. Are these active areas the seat or origin of consciousness; is consciousness generated there; or is this entirely the wrong way of thinking about consciousness?

Cartesian theatre (CT)

Dennett coined the term to describe the common idea that somewhere in the brain or mind, everything comes together and consciousness happens. He argued that most people have rejected standard Cartesian dualism and the homunculus it implies, and yet still think of consciousness in terms of a place or a container. He gives the name Cartesian materialist (CM) to those who claim to be materialists, but still believe in the Cartesian theatre.

In the conversations I tried to draw out whether people think in terms of a CT or not. References to ideas, percepts or information 'entering consciousness', or being 'in consciousness' imply CM, although no one admits to being a CM. Use of theatre and spotlight imagery may also imply

CM but, for example, Baars denies that his theatre is a CT. See figures pp 15, 256.

Dennett, D. C., *Consciousness Explained* (London: Little, Brown & Co., 1991).

change blindness

When a conspicuous feature of a visual scene changes we normally notice. However, if that change occurs during a blink or saccade (large eye movement), or at the moment when a 'mud splash' appears or there is a cut in a film, we do not. This is known as change blindness and may have interesting implications for consciousness. For example, most theories of vision assume that a rich and detailed representation of the world is constructed by the visual system and is then available for conscious experience, or constitutes the contents of consciousness. Change blindness suggests that if trans-saccadic memory is so poor, visual perceptions cannot be detailed representations of the world, and the richness of our visual world may be an illusion.

The most extreme explanations of change blindness, such as that given here by Kevin O'Regan, reject the idea that seeing means building up a representation of the world. See figure p. 167.

Noe, A. (ed.), *Is the Visual World a Grand Illusion?* (Thorverton, Devon: Imprint Academic, 2002).

Chinese nation (China brain)

A thought experiment devised by Ned Block and described in his conversation. He imagines each Chinese person having a radio transmitter/receiver and acting as a neuron in a giant brain. This China brain would then function like an ordinary brain although it would be made of quite different components. Would the whole Chinese nation then be conscious? He assumes not and uses this as an argument against functionalism.

Chinese room

A thought experiment devised by John Searle and described in his conversation. He imagines himself inside a room with lots of Chinese symbols and a rule book telling him how to respond to incoming symbols. He supposes that he would be able to respond appropriately to questions put to him, but without understanding a word of Chinese, and that this refutes strong AI. Said by some to be the most famous challenge to the principles of cognitive science and artificial intelligence, and by others to be a misleading waste of time, there have been hundreds of articles written about it.

Preston, J. and Bishop, M. (eds.), *Views into the Chinese Room: New Essays on Searle and Artificial Intelligence* (Oxford: Clarendon Press, 2002).

dualism

René Descartes (1596–1650) proposed that mind and brain are distinct substances that interact through the pineal gland in the brain, a theory now referred to as Cartesian dualism. Different from this kind of 'substance dualism' is 'property dualism' in which things have both physical and mental properties. Substance dualism is usually compared to monism, the belief that there is only one stuff in the world, whether that is mental (as in idealism) or physical (as in materialism).

Many scientists claim to be materialists but still imply various kinds of dualism in the way they speak about consciousness; for example, talking about the brain 'generating' consciousness (as though it were separate from the brain and its processes), or describing the hard problem in terms of third person facts that are about different kinds of thing from first person experiences. I have tried to draw out these implications in the conversations to find out whether anyone has truly managed to escape from dualism.

emergence

Emergence is usually said to occur when a system exhibits properties that are more than the sum of its parts. A popular example is the wetness of water which cannot be predicted from the properties of hydrogen and oxygen and yet emerges from their combination. However, the concept is hotly disputed within philosophy, and it is not at all clear what people mean when they say that consciousness is an emergent property of brains or of neural activity. For example, they might mean that consciousness is a radically new phenomenon that, once emerged, can act back on the brain that it emerged from, or they might mean only that it is a property which cannot be predicted from the action of single neurons but is in principle understandable once we understand the whole brain.

epiphenomenalism

Traditionally this is the idea that mental events are caused by physical events in the brain but have no effects on that brain. This is a curious and much criticized idea implying that consciousness is something distinct from the brain, but unable to influence it. Unfortunately some people use the term to refer to any view in which consciousness itself has no effects, but this is true of some forms of functionalism in which consciousness itself has no effects because it is not something additional to the physical or functional properties of the brain. This confusion is apparent in some of the conversations.

explanatory gap

The gap in explanation between mind and brain, inner and outer, objective and subjective, or the physical world and consciousness, or the claim that facts about the physical world can never satisfactorily explain facts about consciousness. This is related to the hard problem and what

William James called the great chasm or fathomless abyss. Mysterians such as Colin McGinn or Stephen Pinker say that it can never be bridged; most of my conversationalists believe that it can and will be bridged, but they differ on how. For example the Churchlands, Dennett, and Crick all believe that it will disappear as neuroscience progresses; Hameroff and Penrose believe that it will take a revolution in physics to cross it.

filling in

In each eye we have a blind spot where the optic nerve leaves the back of the eye and yet we do not notice them. The same effect can be demonstrated with artificial scotomas (blind areas) and in people with damage to visual cortex. Is the missing part of the picture filled in? Dennett and O'Regan argue, for different reasons, that it need not be; Gregory and Ramachandran claim that it is.

Ramachandran, V. S. and Blakeslee, S., *Phantoms in the Brain* (London: Fourth Estate, 1998).

first person (approach/method/science/perspective)

The first person perspective is the view from within, how the world seems to me. Few people disagree that this lies at the heart of what we mean by consciousness. The real disputes concern the role of first person methods in a science of consciousness and whether there can be such a thing as first person science. Some people argue that we need special first person methods, while others argue that psychology has always used personal accounts. Some argue for a first person science while others say this is nonsensical because all science must be verifiable by third person data. Another difference concerns the value of disciplines such as meditation or dream work. I asked many of the participants whether they practiced any such disciplines; answers ranged from LaBerge and Varela for whom first person work is essential, to Crick who showed no interest in it.

Varela, F. J. and Shear, J., *The view from within: First person approaches to the study of consciousness*, (Thorverton, Devon: Imprint Academic, 1999).

free will

Said to be the most disputed philosophical issue of all time, free will is the idea that we can act or make choices unconstrained by external circumstances or by an agency such as fate or divine will. Free will is often compared with determinism, in which all events in the world are said to be determined by prior events; a view generally accepted as true among scientists. Incompatibilists claim that free will and determinism cannot be reconciled and therefore if we believe determinism to be true we cannot believe in free will. Compatibilists argue, in various ways, that we can make complex choices that count as having free will even if determinism is true.

Many of my conversationalists expressed versions of compatibilism, including Block, Dennett and Searle; others accepted determinism and claimed that they lived 'as if' they had free will. Some mentioned the experiments by Libet which seem to show that conscious decisions to act come too late to be the cause of apparently free actions. See figure p. 101.

Libet, B. (1985), 'Unconscious cerebral initiative and the role of conscious will in voluntary action', *The Behavioral and Brain Sciences*, **8**, 529–539. See also the many commentaries in the same issue, 539–566, and *BBS*, **10**, 318–321.

functionalism

This is the view that the properties of mental states are constituted by their functional relationships, such as relationships between sensory input and behaviour. It can be contrasted with other attempted solutions to the mind-body problem such as dualism, identity theory, or physicalism. Functionalists believe that if you replicated precisely all the functions of a conscious human brain in a machine then the machine would necessarily be conscious, even if it was made of something quite different from biological neurons. Functionalism has been the mainstream view in cognitive science for some time but is rejected by some philosophers, including Block and Searle.

Global Workspace Theory

A theory based on a cognitive architecture in which currently important information is processed in a global workspace and from there made available to the rest of the system. In this scheme the mind is like a theatre, and consciousness resembles a bright spot on the stage of working memory which is directed by the spotlight of attention, while the rest of the theatre is unconscious. The best known version of GWT was developed by Baars and he explains it in our conversation.

Baars, B. J., *A Cognitive Theory of Consciousness* (Cambridge: Cambridge University Press, 1988).

Hard problem

A term coined by Chalmers in 1994 to refer to the question of how physical processes in the brain give rise to subjective experience; he contrasts it with the 'easy' problems such as understanding perception, memory, learning, or emotions. This is related to the mind-body problem and the explanatory gap, but Chalmers' categorization suggests that when we have solved all the 'easy problems' there will still be something left we do not understand—consciousness or subjective experience. See figure p. 6.

Dualists and mysterians believe that the hard problem is truly hard, while functionalists and identity theorists do not because they claim that once we have understood the functions of the brain or its physical states we will have understood all there is to consciousness. I express various strong versions of the hard problem in the conversations to try to draw out people's beliefs.

Shear, J., *Explaining Consciousness—The Hard problem* (Cambridge, Mass: MIT Press, 1997) (and *Journal of Consciousness Studies* 1995).

identity theory

The identity theory of mind holds that states or processes of the mind are identical to states or processes of the brain. In other words thoughts, ideas, intentions, and experiences are not *correlated with* brain states, or *produced by* brain states, they *are* brain states. This removes any need for dualism but leaves the problem of how such seemingly different things can actually be one and the same. Paul Churchland clearly describes a form of identity theory, although he uses the term 'qualia' which many identity theorists reject.

James–Lange theory of emotion

William James and Carl Lange both proposed, in the nineteenth century, that emotions are the result of physiological responses, such as increased heart rate, muscular tension, or sweating, rather than their cause. As James put it, we feel sorry because we weep and afraid because we tremble, not the other way around.

lucid dream

A dream during which you know, *during the dream*, that it is a dream. Surveys show that 30–40% of people have experienced a lucid dream at least once. Some people have them frequently and a very few people learn to have them at will. Experiencers generally say that everything seems richer and brighter in a lucid dream and that they can control the contents of their dreams when lucid. LaBerge pioneered methods of experimenting with lucid dreams.

Gackenbach, J. and LaBerge, S., *Conscious Mind, Sleeping Brain* (New York: Plenum, 1986).

materialism

The view that the universe consists only of matter, and that all mental phenomena are ultimately explicable in material terms. This is the most popular form of monism. Most scientists are probably materialists.

meme

A unit of cultural transmission; memes include skills, stories, songs, theories, or artefacts, that are copied from person to person. According to the theory of memetics, memes are replicators and culture evolves by the process of variation and selection among memes.

Blackmore, S. J., *The Meme Machine* (Oxford: Oxford University Press, 1999).

monism

Contrasted with dualism, this is the view that there is only one kind of substance in the universe. The two main versions of monism are idealism (everything is mind) and materialism (everything is matter), although there are various forms of neutral monism as well.

neural correlates of consciousness (NCC)

Many scientists are searching for areas or patterns of neural activity that correspond to particular conscious experiences. For example, they may use brain scans or single cell recordings to find out which neurons or brain areas are active when a person reports seeing a particular stimulus or having a particular sensation. For some this approach promises to reveal the causes or location of consciousness in the brain, but to others this idea is misguided. Crick and Ramachandran describe work on the NCC and Metzinger explores the implications of understanding NCCs for society.

NCC is also used to mean the neural correlates of being conscious at all as opposed to being unconscious; as Searle puts it—the difference between a conscious brain and an unconscious brain.

Metzinger, T. (ed.), *Neural Correlates of Consciousness* (Cambridge, Mass: MIT Press, 2000).

neurophenomenology

A marriage between neuroscience and phenomenology, pioneered by Varela and designed to bring together the first person methods of phenomenology with the third person methods of neuroscience.

Varela, F. J. and Shear, J., *The view from within: First person approaches to the study of consciousness* (Thorverton, Devon: Imprint Academic, 1999).

phenomenology (1)

A philosophical tradition founded in the early twentieth century by the German philosopher Husserl and continued by Heidegger, and French philosophers including Merleau-Ponty and Sartre. Phenomenology is based on methods for describing the structures of experience as they present themselves to consciousness, without recourse to theory, deduction, or scientific assumptions. Many attempts have been made to integrate its methods into modern neuroscience, especially Varela's neurophenomenology.

phenomenology (2)

Equivalent to 'subjective experience'. For example people may study the phenomenology of vision, or the phenomenology of pain, meaning the first person experience of vision or pain. Dennett points out that the term originally referred to the pre-theoretical list of the properties of whatever it was people were trying to explain.

qualia (singular: quale)

These are the subjective qualities of any sensory experience, such as the smell of coffee or the blueness of a blue sky. Qualia are often defined in philosophy as being intrinsic properties of experiences (they don't change in relation to each other or anything else). They are sometimes assumed to be private, and ineffable (impossible to communicate to other people). Some philosophers claim that to experience a quale is to know all there is to know about that quale and no one else can know it at all.

There are great debates between philosophers about whether qualia exist or not; for example, the Churchlands say they do and Dennett says they do not. Non-philosophers sometimes use the term very loosely, as a synonym for experience, which confuses the issues.

scan see brain imaging

split brain

In the 1960s some epileptics were treated by cutting the corpus callosum, the bundle of millions of fibres that connects the two halves of the brain. This was done in only the most serious cases to prevent epileptic seizures spreading from one side of the brain to the other. Surprisingly these patients recovered well and showed very few changes in ability or personality, but experiments revealed that the two halves of the brain could communicate independently and, to some extent, operated like separate individuals. Among the interesting questions raised is whether such a person has a split consciousness as well. There are arguments for the split brain person having one, two, or possibly none or many conscious selves. Baars and Searle both entertain the possibility that there are two.

stereoscopic fusion

When two slightly different pictures are shown to each eye the brain can fuse them into a single image which then appears in depth. This is what happens when the two slightly different views from each eye are combined to provide depth cues in normal vision (stereopsis), but it can also be mimicked by the creation of specially designed pairs of images that give rise to strange effects when fused. Examples are stereo pairs in different colours which can be viewed through coloured lenses, and random dot stereograms which at first sight are meaningless but, after some time, fuse to create a 3-D image.

third person see first person

ventral and dorsal streams

The visual system consists of many parallel pathways through which information flows from the eyes to other parts of the brain. Among these, two major streams are known as the ventral and dorsal streams. These

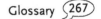

used to be characterized as the 'what' and 'where' systems, but more recently Milner and Goodale have characterized them as systems for perception and visuo-motor control. That is, the ventral stream deals relatively slowly with object recognition while the dorsal stream coordinates fast visually-guided actions. This is relevant to consciousness because fast visuo-motor control seems to happen too quickly to involve consciousness. Some people describe the two systems as though one were conscious and the other not, but Milner and Goodale are careful not to draw this conclusion.

Milner, A. D. and Goodale, M. A., *The Visual Brain in Action* (Oxford: Oxford University Press, 1995).

Index